Thomas Roderick Fraser, Andrew Dewar

The Origin of Creation

Or, the science of matter and force, a new system of natural philosophy

Thomas Roderick Fraser, Andrew Dewar

The Origin of Creation
Or, the science of matter and force, a new system of natural philosophy

ISBN/EAN: 9783337035976

Printed in Europe, USA, Canada, Australia, Japan

Cover: Foto ©Thomas Meinert / pixelio.de

More available books at **www.hansebooks.com**

THE
ORIGIN OF CREATION;

OR

The Science of Matter and Force,

A NEW SYSTEM OF NATURAL PHILOSOPHY,

BY

THOMAS RODERICK FRASER, M. D.

AND

ANDREW DEW......

"The progress of science consists, in the perpetual correction of the errors and falsehoods which preceding minds conceived to be the correct answers they received from nature."
                KINELM CHILLINGLY—LORD BULWER LYTTON.

"It is only by the questioning of received opinions that truth is advanced."
                SHORT STUDIES ON GREAT SUBJECTS—JAMES ANTHONY FROUDE.

"Science must be cultivated for its own sake, for the pure love of truth, rather than for the applause or profit that it brings."
                DR. JOHN TYNDALL.

HALIFAX, N. S.:
PUBLISHED BY THE AUTHORS.
LONDON: LONGMANS, GREEN, READER, AND DYER.
1876.

PREFACE.

UNKNOWN to the world of Science we present ourselves as advocates of the vast undertaking which is, we expect, to revolutionize the whole theory of Natural Science taught and believed in at the present day, and to inaugurate a new system, based upon a natural law, the evidences of which we have discovered, and which we hereby show to be of necessity universal, and therefore capable of explaining all natural phenomena.

This system is not the development of a day, but has been in progress, in various ways, for many years, more particularly since the hidden meaning of MAGNETISM was discovered and applied by us.

The nucleus of the present work appeared some time ago in a weekly periodical, in the form of Essays on Natural Science; and the reception given them, along with the importance and admitted necessity for such a work, has induced us to issue the present volume.

The data upon which our theories on all the subjects touched upon, have been based, are found in a discovery of the nature, classification and properties of the atoms of matter and of the law that governs their action and force, and are proved by practical experiments and under personal

the Gulf Stream, the Calms of the Equator, the coasts of Brazil, California and Mexico, the Mediterranean, the Bay of Fundy, the Hot Sulphur Baths of Salt Lake, the Great Geysers of California, and the mangroves of the Isthmus of Panama.

To our readers generally, let us say, that we desire to be judged

PREFACE.

UNKNOWN to the world of Science we present ourselves as advocates of the vast undertaking which is, we expect, to revolutionize the whole theory of Natural Science taught and believed in at the present day, and to inaugurate a new system, based upon a natural law, the evidences of which we have discovered, and which we hereby show to be of necessity universal, and therefore capable of explaining all natural phenomena.

This system is not the development of a day, but has been in progress, in various ways, for many years, more particularly since the hidden meaning of MAGNETISM was discovered and applied by us.

The nucleus of the present work appeared some time ago in a weekly periodical, in the form of Essays on Natural Science; and the reception given them, along with the importance and admitted necessity for such a work, has induced us to issue the present volume.

~~The data upon which many theories have been built, have, in all the subjects touched upon, been based on personal~~ observations in chemistry, telegraphy and marine diving; in an extensive experience in coal and gold mines; and also while voyaging and travelling along the Gulf Stream, the Calms of the Equator, the coasts of Brazil, California and Mexico, the Mediterranean, the Bay of Fundy, the Hot Sulphur Baths of Salt Lake, the Great Geysers of California, and the mangroves of the Isthmus of Panama.

To our readers generally, let us say, that we desire to be judged only by the light of their faculty of common sense, and their own personal observations in nature without reference to any book whatever, except it may be the Scriptures.

To our possible critics we desire to say that it is useless, for the purpose of convincing us, to attempt to refute our theories by referring to the statements of any man of Science, however eminent, as we recognise no positive authority under God and Nature.

To the many distinguished men now living whose opinions we have ignored, we are personally unknown, and whatever force of language may have been used in refuting their theories, must be attributed to the strength of our convictions on the subject and its commanding importance, and not of course to any unkind feeling to the gentlemen themselves.

We are aware of the imperfect nature of our work, that many unavoidable inaccuracies will present themselves to the careful reader, and that much is comprised in the main part of the work which should appear only as notes; yet we would have these drawbacks excused for the sake of the great truths meant to be conveyed.

The scope of the work also is such—covering as it does facts and systems of Science about which whole libraries have been written—that, owing to our limited space and the necessary condensation, the intent and meaning may sometimes be difficult to apprehend; but we have preferred to publish the Book even in its imperfect condition, in order that we might the sooner obtain the critical suggestions of the scientific world, as a means of rendering it more perfect: for, far from being a work for one man only, there is material to occupy the lives of many scientific men. We have therefore hastened the publication, in order, as intimated, to obtain the assistance of such distinguished men of Science as are still left us, for the rising or progressive men of Natural Science are few, and owing to their cramped ideas, comparatively stationary. Agassiz knew and lamented this fact when he said that we have more than enough of manufacturers of books, men who are mere compilers, who know nothing—of their own knowledge—of the subjects about which they write; while we have few men of patient investigation and research coupled with daring and original thought.

Here then have we lighted our taper to guide the shipwrecked observer who is drowning amidst the swelling seas of opposing theories and systems. Here have we planted our acorn in the already well sown field of science, but whether it will rot in the soil, or the birds of the air will eat it, or the biting frosts will kill it, or whether it will pass unharmed through all these dangers and grow year by year into a mighty oak that shall overtop the forest; time alone will show.

The present systems of science, and theories of accounting for natural phenomena are like to the starry hosts of heaven. Now one startling announcement, with the first flush of youth, passes like the full moon athwart the zenith, dimming all the others; but in half a day it is gone, and it appears next evening only as another speck studded to the starried crown of earth, adding its faint twinkle to the others; yet, after all, there are none capable of illumining the midnight darkness. Many more are like to the evanescent flight of a meteor that does not even leave a stone behind it to tell of its passage. Amidst this host we would also claim a space in which to set our feeble flame, and contribute our quota towards dispelling the gloom of mystery and ignorance; but even this may be denied us.

No greater misfortune can befall a man than to be much in advance of his day and generation. How many hundreds are there probably of such men alive at the present time, who, for want of encouragement, are vainly striving against poverty and misery? While willing enough to raise statues and monuments to them fifty years after they are dead, the world, foolish still and foolish ever, almost invariably refuses to know them while living. When we say, among other things, that MAGNETISM will, long before the present century closes, entirely replace steam as a motive power—for the latter, at the best is only a clumsy, uncertain and dangerous agent to work with—then the tenets which we have advanced are perhaps (without drawing censure on us for egotism) sufficiently ahead of the world's knowledge to wound the vanity of some dozens of professors; to touch the pockets of some thousands whose prosperity would be affected by them; and to render valueless the "loads of learned lumber" in the heads of some millions of bookworms. There is thus sufficient influence—does any one doubt it?—in this interested army to allay the curiosity of the world, and to soothe it back to the even tenor of its way. But, fortunately for us our daily bread does not depend on the acceptance of our theories, and as we watch and wait, and see a few more thousands killed by boiler explosions; a few more thousands drowned by the variation of ships' compasses; a few more millions poisoned by improper medical treatment; a few more fields of coal exhausted, and all our interested professors dead; then perhaps a more intelligent generation will be content to accept the dictation and lessons of Nature.

In the meantime we retain those pleasurable emotions which cannot be taken away from us, the gratification which every writer experiences in unfolding a new idea, the glow of feeling on witnessing for the first time the dawn of a new light on the horizon of knowledge, and the delight in taking home to ones-self a seed of thought garnered from the unfathomable granary of Immensity.

We beg to return thanks to several gentlemen for their kindness in correcting proof, and rendering other valuable assistance.

CONTENTS.

CHAPTER I.
MATTER.

Prof. Grove on Matter.—Locke.—Bishop Berkeley.—Two classes of atoms.—Male and Female Atoms.—Matter on Earth.—Prof. Tyndall on Matter.—Prof. W. A Norton on one kind of Force—Law of repulsion.—*Vestiges of Creation* on Matter.—Fraser's Magazine on Matter.—Analogy between language and two classes of Atoms.......1—5

CHAPTER II.
MATTER AND ITS FORCE.

Atomagnetism.—What Prof. Huxley would like to know.—Matter and motion.—Every Atom a Magnet.—Law of Atoms.—Like attracts Like.—Unlike poles attract.—Atomagnetism the law of attraction and repulsion.—Examples.—Experiments with filings.—How Atoms combine their Polarity.—Herbert Spencer's Philosophy.—His Foundation loosened..6—9

CHAPTER III.
MINERAL LIFE.

Minerals not dead.—Mineral life a low form of vegetable and animal Life.—Iron filings have life.—Compass needle has life.—Philosopher's tree.—Coral.—Candy.—Mineral life.—Atomagnetism.—Atoms of lead, sugar, and coral, Magnets.—Greater always influences the less.—Explanations of Philosopher's Tree.—Cause of beautiful forms in snow flakes..10—11.

CHAPTER IV.
VEGETABLE LIFE.

Origin of Life.—Spontaneous generation.—Sir William Thomson on seed bearing Meteors.—Cornhill Magazine.—Atomagnetism.—No seed required.—Railway Cuttings.—Clover.—How a plant grows without a seed.—Scripture proof for it.—Seeds rot—Hardwood and softwood Forests.—Darwin's "Origin of Species" overthrown.—Thousands of plants in the first creation—New plants with every change of soil and climate.—Present theory of plant life.—How a cell develops—Absurdity of plants breathing.—Why roots and branches spread.—Experiments to prove the reason.—Why a tree does not grow in winter................................12—18.

CHAPTER V.

ORIGIN OF ANIMAL LIFE.

Man afraid to inquire into the origin of life.—Milk and cheese.—Dumas and Agassiz on seeds and eggs.—A cow the mother of maggots.—Insects spontaneously produced.—How animals are produced without an egg.—Excess of vegetable matter forms animals.—Process of creation.—Darwin.—All animals produced not from one but from many.—Agassiz on Men and Monkeys.—One animal may produce a different animal.—Animals, parasites.—Argument against spontaneous generation.—Germ theory.—Pasteur.—Child.—Lamarck.—Canned meats.—Why ice and salt preserve meats.—The formation of germs.—Tyndall on respirators.—Spontaneous fish.—Agassiz on Special Creation.—Origin of lowest organisms.—Mr. Charlton Bastian...19—29

CHAPTER VI.

APPETITE, OR INCIPIENT MIND.

Darwin thinks development of Mind a hopeless inquiry.—We explain it.—Appetite the lowest form of Mind in Animals.—Spontaneous Insects eating immediately.—What is Appetite?—The Atomic Law of Like to Like.—Mind and Life, Properties of Matter.—Vegetable Appetite.—A Seal's Appetite.—A Calf's Appetite.—Why it does not eat bricks and stones.—A Baby's Appetite.—Appetite for Tomatoes.—Superiority of a Brute's Appetite over Man's..30—32.

CHAPTER VII.

INSTINCT, OR ANIMAL MIND.

Instinct a higher phase of Mind.—Frank Buckland.—Why a Chicken knew a Gentleman was not its Mother.—Sparrows require no Teaching.—Foreknowledge of Bees and Beavers.—Important Fact.—Animal's Mind Perfect.—Never Progresses.—Man always Progressing.—Difference between man and Beast.—Man two minds.—Animals one.—Animals no Soul.—Mind returns to Earth.—Their Mind all Nature.—Animals Perfect on separation from the Parent.—Answer to Frank Buckland's questioner................33—36.

CHAPTER VIII.

MAN'S ANIMAL AND SPIRITUAL MIND.

Schelling and Hegel on Nature as "petrified intelligence."—Hope on "Origin and Prospects of Man."—Matter without properties.—Mind a property of matter.—No limit to the properties of matter.—Brutes have one mind, man two minds.—Animal and Divine.—Agassiz on two minds.—Why Man's animal mind degenerated.—Man should distrust man.—Manner in which man's mind is formed.—From food.—Difference between animal mind and Divine.—Situation of the mind.—Of Memory.—Brain a picture gallery.—Difference between man's mind and the brutes....................37—42.

CHAPTER IX.

CHEMICAL ACTION.—STEAM BOILER EXPLOSIONS.

A knowledge of chemical action requisite.—Nothing known about it by scientific writers.—Prof. Grove.—Chemical action only one form of atomagnetism.—Great separater.—Attraction the great builder.—Repulsion the great designer.—Chemical action the great destroyer.—How sugar dissolves in water.—How a nail dissolves.—Concentrated acid not so good a dissolver as diluted acid.—Soda powder, Sulphuric acid.—Amusement for speculative philosophers.—How water evaporates.—No latent moisture in the atmosphere.—No latent dryness in the sea.—STEAM BOILER EXPLOSIONS.—Facts connected with explosions.—The materials dealt with.—The manufacture of hydrogen gas.—What the United States Commissioners on explosions have discovered.—How explosion takes place.—Not by pressure.—Mingling of gases.—Prevention..........43—51.

CHAPTER X.

HEAT.

Heat the result of chemical action between certain classes of atoms.—Dynamical theory of heat.—Motion.—Tyndall on heat.—Christopher Columbus and his followers.—No ambition among scientific men.—Heat produced in three ways.—Natural heat.—Combustion.—Friction.—Ice melting.—Hot springs and Geysers of California.—Volcanoes caused by chemical action.—Why coal burns.—Poker experiment.—Conductive power of heat.—Tyndall's experiments.—Laboratory experiments incorrect.—Atomic action likened to a gossamer thread.—How the Crusade against the present system of science will be conducted.—Ruskin's Crusade against Renaissance Painting, and Architecture.—Grove and Lardner........52—61.

CHAPTER XI.

LIGHT.

Light caused similarly to heat.—Propagated differently.—Three divisions.—Light without heat.—Light with heat.—Propagated light.—Auroras explained.—Phosphorescence.—Tyndall refuted on molecular motion.—Fire-flies.—Lighting gas by the finger.—Auroras from trees.—Candle a guide to light.—Four things required to be looked at.—The flame.—The heat.—The light.—And light as an object.—All light, reflection.—The undulating theory disputed.—Light instantaneous.—Light cannot travel half a mile.—Sight travels 286,000 miles a second.—Flame not seen in daylight.—Astronomical fallacy of star-light.—Undulation follows Emission into oblivion.—Tyndall's security for the continued acceptance of the Undulatory theory, overthrown...............62—68.

CHAPTER XII.

THE SUN AND SUNLIGHT.

Professors Thomson, and Tait, on the Sun.—The Sun a huge furnace.—Herschell on the waste heat of the Sun.—Temperature of space.—Our view of the universe.—The solar system an inhabitant of it.—The Sun a stomach.—The atmospheres, the flesh and bones of the solar system.—Movements regulated by Magnetism.—The Sun an inhabited world.—How Sunlight is caused by magnetism.—The Sun, Earth, and Planets, Magnetic batteries.—Sun the main battery and head office.—Planets telegraph stations.—Sunlight caused in a similar way to the spark at the poles of a battery.—The "Journey to the Sun." .... 69—76.

CHAPTER XIII.

COLOUR.

Undulation theory of colour.—What is the force which governs colour.—Primary causes overlooked as usual.—Great display of Arithmetic.—Looseness in Science.—When we will freeze to death.—Portland Scientific Convention.—Tyndall on the vibratory theory.—474,439,680,000,000 red waves a second.—This theory questioned.—No colour on the Earth.—Herschell.—Helmholtz.—Science like a voyage of discovery.—We introduce the atomagnetic theory of colour.—Colour a property of matter.—Colours of mineral flames.—Why is the sky blue?—Tyndall's "Scientific use of the Imagination."—The setting sun red.—The hills purple .... 77—83.

CHAPTER XIV.

ELECTRICITY.

All light is Electricity.—Greatest Scientific delusion of the day.—Magnetism and Electricity essentially different.—Quotations to show how little is known about either.—Dr. Thomson.—Parker's School Book of Philosophy.—Sir Wm. Thomson on Electricity flowing.—Prof. Tyndall also confesses ignorance.—Prof. Grove.—Prescott's History.—Electric spark, what composed of.—No combustion without a mixture of the two classes of matter.—The cause of lightning .... 84—89.

CHAPTER XV.

MAGNETISM.

Explanation chapter.—To show difference between Magnetism and Electricity.—Profs. Grove and Faraday.—Electricity not a force at all.—Arrangement of a galvanic battery.—How telegraphing is accomplished.—Telegraph worked by grass.—A few facts about magnetism.—Well known and not generally known.—Faraday's misfortune.—Born too soon.—The "Magnetic curves" explained.—Tyndall astray again.—Polarity of iron railings.—How the polarity of magnetism changes with position.—Sir Isaac Newton's apple.—The law of gravitation upset.—How magnetism is a weight, and how it affects weight.—What Newton wished to discover.—The cause of deviation in iron ships .... 90—101.

CHAPTER XVI.

SOUND.

Difficult problem in Science.—Prof. Tyndall's explanation not satisfactory.—Sound vibrations and light vibrations.—Sound generates heat.—How long fifty organs would take to heat St. Paul's Cathedral.—Sound in summer and winter.—How we hear fifty sounds at the same time.—Echoes.—New theory of Sound.—A sympathy between the mineral atoms of matter.—Iron a better conductor than wood.—If a man has sympathy why should not an atom?—Dancing flames.—Tyndall's new theory of Sound.—Experiments at the South Foreland, England.—Vapour in layers.................102—106

CHAPTER XVII.

WATER AND RAIN.

Fire not so powerful as water.—Water in granite.—Herschell on Rain.—Rain caused by chemical action in the atmosphere.—The Rain guage.—Rain forms in the lower atmosphere.—Proctor and Kamtz on the reason why.—Rain shot out from clouds.—Herschell on Rain storms.—Climate of North America changing.—Egypt cultivating the Palm for Rain.—Forests and vegetation cause Rain.—Herchell's reason why, a failure.—Drainage said to be bad.—Chicago, St. Louis, once unhealthy.—Why.—No large city unhealthy.—No air in water.—Fishes gills used for filtering food, not for breathing.—The air they need produced from digestion.—Can we produce or bring down Rain?—Great battles in America were followed by Rain.—The cause.—.................................107—112.

CHAPTER XVIII.

DEW.

Chambers Journal.—Baptista Porta nearly discovered the true theory of dew.—Thought dew was condensed from air.—Aristotle thought it was condensed from vapour.—Muschenbrook kept back Meteorology one hundred years.—Great discoveries often foiled by the stupidity of the world.—Dr. Wells said to be the discoverer of the true dew theory.—The radiation of heat, the basis.—The cause of moonblindness.—Dew forms most readily on vegetation.—Arguments against radiation.—Observations with wool packs.—Position everything.—Calm and clear evenings essential.—Dew is water.—Produced in a similar way.—The cause of fog and hoar frost.—Hoar frost spears of ice..119—123.

CHAPTER XIX.

THE ATMOSPHERE AND STORMS.

Atmosphere said to be composed of oxygen and nitrogen.—An impossibility.—Air in no two places the same.—Balloon explorations.—Guy Lussac.—Everything with life has an atmosphere.—The atmosphere of the African.—Impossible to get rid of it.—The earth a living body.—Has an Atmosphere composed of its own materials.—The

Atmosphere composed of hundreds of different compounds of materials.—STORMS: Sir John Herschell and Prof. Rogers on Storms.—Magnetic curves from the poles of the earth, the cause of wind and storms.—Cause of Equatorial Calms.—Maury on cyclones.—Description of a so-called Circular Storm.—Hints for Weather Prophets...124—132

CHAPTER XX.

ANIMAL, VEGETABLE, AND MINERAL FOOD.

Nothing so much to do with our discomforts as food.—The body a machine.—Professor Lyon Playfair on Food.—Liebig's classes of food.—Flesh formers.—Heat givers, and mineral ingredients.—Knows nothing of the action of the last class.—Contradictions.—Experiments by scientific men always conducted too loosely.—Animal food only concentrated vegetable matter.—English Navvies and Arabs.—Sepoys and Ghoorkas.—How much an Esquimaux eats, according to Sir John Ross.—Canadian Indians and salt.—Criminals in Holland.—The Scotch and Indigestion.—The action of minerals in the body...133—138.

CHAPTER XXI.

COAL.

Found to be of vegetable origin.—Prof. Rogers on Coal.—Statements faulty.—Unacquainted with natural law.—Rogers' theory.—Grew in a swamp.—Soaked with mineral oils.—Baked by the earth's internal fire.—A forest makes half an inch of coal.—A tree said to absorb carbon.—Incorrect.—Sir Henry De la Beche and his calculations.—Fallacies about carbon.—How carbon and hydrogen came into the coal.—Our theory of coal.—Prairies.—Charcoal in the seams.—Nova Scotia mines.—Inundations.—No internal fire.—No Baking.—The whole process one of petrifaction.—Coal inexhaustible..139—144.

CHAPTER XXII.

HOW CORAL GROWS.

Strange Chapter.—Coral insects unworthy of notice.—Misplaced eulogy.—Theories of Coral growth.—The insect monument and tomb.—Not found below thirty fathoms.—Coral found a mile and a half deep.—Coral on the Isthmus of Panama, not made by insects.—The Coral insect a parasite merely.—The cochineal.—How Coral grows.—Millions feeding from one mouth.—Coral grows by budding.—Agassiz on Florida reefs, and arguments against Darwin.—Darwin's curious theories on Coral reefs.—Sir John Herschell.—How Coral commences to grow.—The true theory of reefs.—How a gap in a reef was filled.—Coral merely the home of the insect...145—152.

CHAPTER XXIII.

VOLCANOES AND EARTHQUAKES.

Another popular fallacy.—The earth's internal fire.—Dr. Mayer's theory.—Dr. Tyndall opposed to it.—Dr. Mayer's dogged assertion.—Selfishness of men of science.—Herschell on Volcanoes.—The earth and an egg.—Objection to Herschell's theory.—Explanation of Volcanoes.—Why Volcanoes become extinct.—Coal gas.—Mount St. Helena and Sulphur Springs.—Prof. Mallet on Water and Volcanoes.—Cause of Earthquakes.—Prevention of Earthquakes.—Oil boring in Pennsylvania.—Herschell's extraordinary theory of Earthquakes.—What he knew of chemical action in the interior.—The necessity for scientific men not taking anything for granted. 153—160.

CHAPTER XXIV.

THE TIDES.

The regularity of the Tides.—The influence of the new and full moon on the Tides.—There must be one grand cause of the Tides.—This is pressure, not attraction.—Cause of variation in the Tides by the position of the moon.—Formation of the Land.—Winds.—Lardner's theory of the Tides.—Its fallacy shewn.—The earth ought to be approaching the moon.—Facts to be remembered.—The Plane of the Ecliptic.—The effect of pressure on the atmosphere.—The Tides caused by pressure in passing the Plane of the Ecliptic.—The moon's atmosphere.—The Tide in the Mediterranean.—The Bay of Fundy Tides seventy feet high.—Ram Pasture.—Rise of two feet in three miles.—The repelling forces control the Tides.... 161—166.

CHAPTER XXV.

THE GULF STREAM AND DEEP SEA CURRENTS.

The cause of the Gulf Stream.—Dr. Carpenter's theory.—Oceanic Circulation.—Experiment with glass trough.—No comparison.—Strength of Polar Currents.—Channel between Faroe Islands and Shetland.—Dr. Wyville Thompson differs from Dr. Carpenter.—Reciprocal circulation of Water and Air.—Beautiful theory of Atmospheric Circulation overlooked by Dr. Carpenter.—What causes the cold deep waters.......................... 167—172

CHAPTER XXVI.

COMETS.

Very little known about Comets.—Facts about them.—Jupiter's influence on them.—Comet of 1680.—Herschell's description of it.—The movements of a Comet different from a Planet.—All the heavenly bodies, Magnets.—The motions of Comets explained on this theory.—How Comets are made periodic.—Encke's and Biela's Comets.—The atmospheres of Comets.—Their tails.—Their purpose.—Are they inhabited?.................................. 173—177

CHAPTER XXVII.
METEORS.

Strange theories regarding them.—Sir Wm. Thomson's.—Seed bearing Meteors.—Prof. Newton on November Meteors.—No orbit of Meteors.—Meteors caused by pressure and reciprocation.—Dr. Sorby the microscopist on Meteors.—Prof. Graham on the Leonarto Meteor.—The great November showers caused by a Comet.—The yearly and ordinary Meteors caused by pressure..................178—181.

CHAPTER XXVIII.
AURORA BOREALIS.

Visible at both poles.—Mairan on the extent of the Sun's atmosphere.—Lardner on Auroras.—M. Biot on Polar Volcanoes.—Distance of Auroras.—Seen by Aeronauts below them.—Facts.—Caused by mineral emanations from the Polar Latitudes.—How they affect the compasses.—Why seen on Calm evenings.—Dew.—Cause of colours.—Similarity between Auroras and Meteors............182—185

CHAPTER XXIX.
MEDICINE; OR, THE LIFE ACTION OF THE BODY, AND THE CAUSE AND CURE OF DISEASE.

A revolution in medicine.—The cause of disease unknown.—Incurable diseases.—Not creditable to the profession.—What are our bodies composed of?—What keeps up life in us?—What is blood?—How is blood formed?—How is the material we eat transformed into blood?—What causes and keeps up the circulation of the blood?—What is life?—The magnetic action of the body.—The function of the blood.—How the waste from the body is thrown off.—Hot water.—Purging—Emetics.—The body compared to a fire.—Indigestion.—Consumption, its cause and cure.—Fevers....186—195.

CHAPTER XXX.
ATOMAGNETISM AND RELIGION.

Religion not affected by Atomagnetism.—The inherent life in atoms and the spontaneous development of the mind, seem grand arguments for the Materialist.—The movements of Planets and Comets.—The great machinery of the universe.—What need of a God?—Man fancies himself a Monarch.—No animal intelligence his superior.—Only a parasite.—Chained to the earth.—On a level with his dog.—Matter existed without properties.—Who endowed it with them?—Divine mind of man.—Magnetism not God.—How simple, miracles must be to Him who formed and holds the key of natural law.—Insignificance of man ....196—199.

MEN OF SCIENCE QUOTED OR REFERRED TO.

Aristotle.
Agassiz, Louis

Bacon, Lord
Bastian, H. Charlton
Beche, Sir Henry de la
Berkeley, Bishop
Biot, M.
Brewer, Dr.
Buckland, Frank

Carpenter, Dr. W. B.
Child, Dr.
Coulomb, M. A. C. De
Crosse, Mr.

Davy, Sir Humphrey
Darwin, Dr.
Descartes.
Dumas, Prof.

Faraday, Michael

Grove, Prof. W. R.
Graham, Prof.

Hegel.
Helmholtz, Prof. H. L. M.
Herschell, Sir J. F. W.
Higgins, W. M.
Huxley, Prof.
Hope.

Kamtz.

Lamarck.
Laplace.
Lardner, Dr. D.
Liebig, J. Von

Lussac, Guy
Lyell, Sir Charles

Mairan.
Mallet, Prof.
Maury, Dr. M. F.
Maury, Prof. Thompson B.
Maxwell, Prof.
Mayer, Dr.
Muschenbrook.

Newton, Sir Isaac
Newton, Prof.
Nichols, Prof.
Norton, Prof. W. A.

Pasteur, M.
Playfair, Prof. Lyon
Porta, Baptista
Proctor, Prof. R. A.

Rive, Prof. A. De la
Rogers, Prof.
Ross, Sir John

Schelling
Schiaparelli
Sorby, Dr.

Tait, Prof. P. G.
Thomson, Sir William
Thompson, Dr. Wyville
Thomson, Dr. Thos.
Tyndall, Prof. John

Wells, Dr.

Young, Dr. Thos.

LIST OF NEW DISCOVERIES IN SCIENCE.

The duality of atoms.
The properties and force of matter.
The cause of life.
The source of mind.
The cause of chemical action.
The cause of sunlight.
The cause of variation in ships' compasses.
The cause of boiler explosions.
The cause of winds and storms.
The process of digestion.
The cause of the tides.
That magnetism is weight, and supersedes gravitation.
That coral is a semi-mineral growth and not the work of insects.
The cause of meteors.
The cause of auroras.
The cause of the circulation of the blood.
*That hydrogen gas has the properties, not only of metals, but of minerals.
That oxygen gas has the properties of vegetable matter.

*We are aware that Prof. Graham has the credit of the discovery, that hydrogen was metallic; but as we published the announcement of the same fact in the Phrenological Journal, in 1863, Prof. Graham's alleged discovery was thus anticipated by some years.

INTRODUCTION.

To PROPOUND a new system of Natural Philosophy in this age of enlightenment and great men which shall overthrow the cherished theories of centuries, as well as those of later date, is a difficult, and, we suspect, a thankless task; but it is one which, in the interest of truth, the progress of knowledge, and the eradication of sensationalism in Science, we feel compelled to undertake.

If it is acknowledged that no one knows the composition of matter; that the force of matter is unknown; that chemical action is a mystery; that life and mind are inexplicable; that electricity and magnetism are forces but partially understood, and that over all natural phenomena there hangs a veil of mystery: then, if our most voluminous writers on science mean what they say in their reverence for the truth and their endeavours after its acceptance, we, who offer an explanation of all these mysteries, should receive encouragement and assistance on every hand.

It must be morever the desire of every intelligent man, outside of scientific circles, that some more definite system of science should be adopted than that in which belief is generally placed. People are beginning to tire of the extraordinary theories regarding the sun, moon, and stars, which are successively being advanced, and which intelligent men are compelled to read if not to accept—if they would keep pace with what is called the progress of knowledge—despite, too, their own doubts and convictions of error.

Of course it is the last result that the lower grades of scientific men generally would try to bring about. It would lower them from the proud position of heroic poets—gifted with an illimitable imagination, and furnished with an unbounded license to terrify mankind—to the level of common mortals like ourselves. No more would their names be heralded by the journals of the world accompanied by some brain-whirling paragraph, and unless they were really possessed of more intellectual power than their brethren generally, their names would never be heard of.

For instance, see what happened at a late meeting of the American Scientific Association at Portland, U.S.; and although Americans, they only imitate the theories advanced by European *savants*. Five papers were read apparently to horrify the audience; each having as its grand conclusion, the extinction of life on the earth. We beg to be excused for giving their names such prominence. Prof. Young said, in substance, that the sun was being gradually muffled by a peculiar rain falling on it, which formed a crust that would eventually exclude all light and heat, so that a return to original chaos would be the inevitable consequence. He forbore giving the exact date of the catastrophe.

Gen. J. G. Barnard came more to the point. He said our earth is only a fire-bubble with a very thin crust, so that we are liable to explode at any moment. As soon, therefore, as we hear of any telegraphic report of a volcano in eruption, or see any heavy meteor or comet dashing towards the earth, then, too, we may listen for the sound of the last trumpet.

Mr. H. F. Walling read an essay on the "Dissipation of Energy" in which he stated that the sun was losing its heat so rapidly that there will be a slowing of the machinery of the universe, until stagnation culminates in a total extinction of life. No date given.

Prof. Franklin B. Hough foretold a perpetual drought in consequence of clearing the forests. The result will be a universal famine, and the world will be depopulated by starvation. No date given.

The last paper was by Dr. Le Conte, the new President of the Association, and he foretold there would be such an alarming increase of insects, that all vegetation would be destroyed, and finally starving and helpless man himself would be eaten. No date.

"All of which," says an American paper, "argues an early dropping of the curtain upon the fleeting show of life."

Is it at all possible that a system of science can be true which permits such an outcrop of startling absurdities, no speculation being too ridiculous to be issued, while the inventor of the most terrifying announcement becomes the most celebrated man of the time? The wonder, too, is that such illustrious men as Herschell should be led away by the prevailing weakness. It is to be hoped that it was only as a joke when he said that some of the spots on the sun, 600 miles long by 300 miles wide, might be *living creatures;* but Prof. Proctor in his New York Lectures seemed to quote it as a statement made in earnest. Prof. Proctor himself indulges in some visionary dreams regarding the exhaustion of the solar heat and the aspect of the inhabitants of Saturn, which are but a shade removed from the absurd. There is nothing so correct and invariable as nature in all her laws, and thus no study should be so free from sensationalism as Natural Science. The very semblance of it ought to be an abomination to the true student.

While, therefore, it is admitted that the established system of science—if it may be called a system—is incorrect: for it is not only we who say so, but every great physicist from Newton downwards has acknowledged that there was something lacking: yet Science has collected a vast array of valuable facts which only want an assorter. They are like the multitude of objects sent to an Universal Exhibition, but the building in which they are to be displayed has still to be erected. Or they are like the hieroglyphics on an Egyptian tomb waiting for a Rawlinson to interpret them; we hope that they will not be like the Aztec characters on the Central American ruins which have waited, and are, as far as appearances go, likely to wait in vain for an interpreter.

Every river has a source; every tree has a root; every building has a foundation; but it is confessed by all men that Natural Science at present has no source, no root, no foundation upon which to stand. The different theories are like Arabian streams which are lost in the sands of the desert; or northern lights which are never seen twice in the same place. A pole or guiding star has continually been sought, but, like the North West Passage, it has eluded all search, and many brave men have died in the pursuit. Still, we do not blame explorers for not finding it. Man is not omniscient, and even a Franklin may fail to reach the North Pole, or a Buckle may die when his work is half done. But every man should be honoured for the work he has accomplished, if performed conscientiously and with originality, according to his lights. We do not consign the heathen to hell because they do not happen to know there is a true God; but we certainly condemn those men who, knowing evil, continue to preach and practice that evil. So while we use all forbearance to those who have been educated in a false system of science, we would unmercifully scathe those bigoted pedants who would spurn the truth when it is offered them; who would rather continue to teach false doctrine, knowing it to be false, than condescend to learn a science which was true.

We feel that we have been censorious and that we have perhaps condemned the innocent with the guilty. We know that we have spoken irreverently of names that do honour to our race. But it is a necessity almost forced on us by the nature of the work. Our object is not so much to show where Newton, Herschell, Agassiz, Tyndall, Thomson and others are right, but where they are wrong; not to praise them but to condemn them where they deserve it; for it must be admitted that there are few who do not deserve censure. If our object was to praise alone, nothing would afford us greater pleasure, and we are conscious that we could find much in the life and work of these men overlooked by the superficial flatterer whereby to exalt and do them honor; but praising a man who is successful, is a work which is already too well done by an army of parasites and sycophants all the world over; an army which is also ready at the same time to tear to shreds the reputation of a genius who stumbles in his life struggle. There is a praise which degenerates into fulsomeness, and a worship which degenerates into toadyism; and while the refusal to give merit to whom it is due, is bad, the over-praise of a man leading to the general belief that he is an infallible authority,—a prerogative conceded only to nature herself—is infinitely worse.

The great bane of cultured progress in the present, if not in all centuries, has been the worship of authority. If the Pope says the sun goes round the earth, then Galileo must believe him. If Sir Isaac Newton says gravitation is the universal law of earth, then Herschell will not question the fact; and strange to say—although many knew the discrepancies of that law, and the many exhibitions of force which it was unable to explain,—no man up to the present time has had the manliness to speak against gravitation. Scientific men seem to have gone on the principle that a law, although defective, is better than no law at all; just as many nations to their cost have said that a bad government is better than no government. But as a bloody revolution is the inevitable destiny of such a people, and good government, easily to have been obtained if sought for in time, is at length only attained by a sacrifice of life which blackens the page of history; so it will be in science, for the tree of knowledge has been so overgrown and entwined with creepers, that its growth has been choked and stunted, its fair proportions destroyed, and its vitality threatened. In order, therefore, that it may again branch out in all its beauty, it will have to be severed at the roots.

This worship of authority has poisoned the streams of other branches of knowledge, for Architecture slept at the Reformation, and until lately a blind copyism of Grecian, Roman, and in due sequence all the types of the Gothic styles prevailed, so that artistic feeling was almost quenched in the architect who would be popular. So also was it in Painting, where Cupids, Venuses, Madonnas, and artificial landscapes were the main staple of art.

Our Theologians also are still continually trammelled and led into trouble by quoting authority two or three hundred years old; while Doctors allow their thousands of patients to die annually through the like blind worship.

It will be said that some authority must be acknowledged, else there would be neither science or government. Unquestionably. But because authority is in power, it does not follow that it should remain unquestioned. All law except nature's is fallible, and can only be kept right by a continual examination. It is the cashier in whom implicit confidence is placed that usually embezzles the funds of the Bank. So if we would have authority—and it is a necessity—it must be one that undergoes a continual scrutiny, and answers every interrogation promptly. As long as it does so, then reverence it; but once it fails, look out for another more sure. Do not try to prop up a fallacy. The ruin is only the greater when it does fall.

It may be said that we should have accepted at least some men as authorities. In many things we certainly do, but where their theories conflict with obvious truth, then we throw them at once to the winds. Galileo, Sir Isaac Newton, Bacon, and Sir Humphrey Davy were all celebrated men in their day. They were all students of nature, and each, as was said by Newton, picked up but a pebble of truth from an inexhaustible shore, knowing that there were many more yet to be discovered. Far be it from us, therefore, to detract anything from the honour they deserved, and the glory they earned. But knowledge is no law of the Medes and Persians which changeth not. The dullest school boy may now know what these men would have given worlds to understand. Is it then worthy of such intelligence as we are possessed of, or worthy of this grand century of thought and discovery, to have so little confidence in ourselves as to place implicit belief in theories which these men,—great though they were,—in their imperfect knowledge laid down, while facts are every day being brought to our notice antagonistic to them? We believe that such men were above a paltry adulation. While some philosophers blind themselves to consequences which the recognition of facts entail, and would rather believe that the phenomena never occurred than that Newton should be wrong; we believe that Newton's, or any other sensible man's last wish would be, that anything he had said should stay the progress of truth. Besides, in such false humility we do injustice to ourselves and are ungrateful to the age we live in.

The sun of knowledge is ever brightening as the years roll on, making the hidden places clearer and the difficult paths easy of travel. But this light has been unrecognised, and this sun overlooked in preference to those twinkling stars, which in their own day and in their own system shone as suns with brightest effulgence, but which to us as the years glide by, are now no more than brilliant and beautiful gems in the sparkling galaxy of the past.

The present work we believe to be the first attempt that has yet been made to arrange the sciences under one common head, and to show how they are all governed by one and the same law. How far we have been successful, our readers will decide. We suspect that there may be something observed in every chapter to startle the ordinary scientific student, but we offer no opinion which cannot be proved correct by a reference to the operations of nature, since we have been guided entirely by her teachings.

To detail fully the manner in which we arrived at the principles of our theory and the experiments performed, would occupy too much space, we therefore give the following brief statement :—

Finding all theories of Natural Science to be conflicting and unsatisfactory even to Scientific men, we laid them aside and referred to Nature for explanations of her working. By tracing every phenomenon to its origin we found all phenomena to spring from one and the same source. That is, the variety of Natural phenomena are not—as is generally supposed—caused by a variety of forces and a variety of laws, but result from the varied compounds, conditions and positions into which matter may be placed, operated on by its own inherent force under *one* law that controls the whole. The theory is then as follows :—

Matter is composed of two classes of atoms, mineral and vegetable; or, as they are often called throughout the work, Hydrogen and Oxygen. Every atom is a magnet having polarity. ~~Like atoms attract. Like poles repel, and unlike~~

Atoms and their magnetic force being inseparable, their dual law properly reads thus : Like atoms attract and repel each its like and those of its class only, the greater body attracting or reversing the polarity of the smaller. Like poles repel—thus dissolving matter, and unlike poles attract, thus building up or forming material substances, so by nature's laws all work together in harmony.

That the law is correct and complete we have no doubt; but we do not assert that we are right in the interpretation of every detail of the law, as exhibited in all the variations of natural phenomena, for infinite wisdom is not attained by man.

Thus while all other systems fail to give a reliable explanation of the simplest natural phenomenon, we are confident that worthy of this grand century of thought and discovery, to have so little confidence in ourselves as to place implicit belief in theories which these men,—great though they were,—in their imperfect knowledge laid down, while facts are every day being brought to our notice antagonistic to them? We believe that such men were above a paltry adulation. While some philosophers blind themselves to consequences which the recognition of facts entail, and would rather believe that the phenomena never occurred than that Newton should be wrong; we believe that Newton's, or any other sensible man's last wish would be, that anything he had said should stay the progress of truth. Besides, in such false humility we do injustice to ourselves and are ungrateful to the age we live in.

The sun of knowledge is ever brightening as the years roll on, making the hidden places clearer and the difficult paths easy of travel. But this light has been unrecognised, and this sun overlooked in preference to those twinkling stars, which in their own day and in their own system shone as suns with brightest effulgence, but which to us as the years glide by, are now no more than brilliant and beautiful gems in the sparkling galaxy of the past.

The present work we believe to be the ——
yet been made ——
and to
law.
decide. ——ng observed in every ch —— ordinary scientific student, but we offer no opinion which cannot be proved correct by a reference to the operations of nature, since we have been guided entirely by her teachings.

INTRODUCTION.

To detail fully the manner in which we arrived at the principles of our theory and the experiments performed, would occupy too much space, we therefore give the following brief statement :—

Finding all theories of Natural Science to be conflicting and unsatisfactory even to Scientific men, we laid them aside and referred to Nature for explanations of her working. By tracing every phenomenon to its origin we found all phenomena to spring from one and the same source. That is, the variety of Natural phenomena are not—as is generally supposed—caused by a variety of forces and a variety of laws, but result from the varied compounds, conditions and positions into which matter may be placed, operated on by its own inherent force under *one* law that controls the whole. The theory is then as follows :—

Matter is composed of two classes of atoms, mineral and vegetable; or, as they are often called throughout the work, Hydrogen and Oxygen. Every atom is a magnet having polarity. ~~Like atoms attract. Like poles repel, and unlike poles attract~~.

To this universal law of the attraction of like atoms and the repulsion of like poles by their inherent magnetic force, we have applied the term ATOMAGNETISM.

That the law is correct and complete we have no doubt; but we do not assert that we are right in the interpretation of every detail of the law, as exhibited in all the variations of natural phenomena, for infinite wisdom is not attained by man.

Thus while all other systems fail to give a reliable explanation of the simplest natural phenomenon, we are confident that the Atomagnetic will never fail. That it will always give the right answer will depend upon the mind of the man who asks the question, for it is not always the law which answers a question, but, unfortunately for science, the fallible mind of man himself assumes the responsibility. *Thus while we have discovered the law we do not claim to be the infallible interpreters of it.*

People generally judge of a scientific theory by its practical value and usefulness. Looking at it even in this light (which is disclaimed by modern scientific writers for very obvious reasons) Atomagnetism is bound to be of immense permanent value. We expect to save the lives of thousands by showing the cause of boiler explosions, by explaining the variation of ship's compasses, and by showing the cause of life; also how our bodies are kept in life, and how sick people should be treated. But we suspect that the most valuable practical discovery we have made will be that of the nature and action of *magnetism*, which shows that it may be brought under control, and made to subserve man's will and work in every department of labour. It is strange, that if magnetism is the only force which nature employs to accomplish her mighty work in all her actions and movements, that man cannot make it propel his tiny engines and machinery. Without going into particulars we may say that we have made a machine and propelled it by magnetism. We proceeded with our experiments far enough to satisfy us that there was no limit to the power that might be obtained; and this power could be gained without smoke, without fire, without danger of explosion, and at a cost of only a thousandth part of that of steam. It seems like a dream but it is a dream that will yet be realised.

In conclusion, the realm of science, we have often thought, resembles the vast expanse of ocean that lies glistening in the morning sun; not a breath of air disturbs the glassy surface; the vessels lie lazily without movement or sign of life, and the spars are all clearly reflected in the water. Anon as we cast our eyes out to sea, we perceive a shade darkening the horizon. On it comes gradually widening and expanding till it spreads like the shadow of an eclipse over the sea, throwing the water into foam, spreading a mantle of blue where all was white, and converting the lazy boats into birds of flight, so that each ship could now enter on its voyage and go where it willed, while previously it had to drift with tide and current. The ocean is the sea of all knowledge; the wind is the guiding law which leads to the sources and ends of all truth; and the vessels are scientific theories which drift about with the tide. Sometimes the name of a man may urge a theory along and give it prominence, just as we sometimes see a tug boat towing a vessel out of a harbor. But the steamer cannot always be with it, and after the vessel is alone it is still at the mercy of the tide. The ships may also make a great display and noise, they may hoist flags and fire off cannon, they may challenge respect and attract considerable attention, but they are only becalmed after all. Which, then, of all the theories before the world, is that favourable wind?

THE ORIGIN OF CREATION.

CHAPTER I.

MATTER.

Prof. Grove on Matter.—Locke.—Bishop Berkeley.—Two classes of Atoms.—Male and Female Atoms.—Matter on Earth.—Prof. Tyndall on Matter.—Prof. A. Norton on one kind of Force.—Law of repulsion.—Vestiges of Creation on Matter.—Fraser's Magazine on Matter.—Analogy between language and two classes of Atoms.

PROF. GROVE says:—" Probably man will never know the ultimate structure of Matter, or the minutiæ of molecular action; indeed it is scarcely conceivable that the mind of man can ever attain to this knowledge." Locke also affirms that we know nothing of substance of any kind. Bishop Berkeley also said to the Materialists:—" You tell me that all the phenomena of nature are resolvable into matter and its affections. I assent to your statement, and now I put you the further question, What is Matter ?" This was a puzzler for them; but, with all due deference to such high authorities, we profess to have discovered not only what matter is composed of, but its properties and law as well. The consequence of this has been the discovery of other natural laws, or variations of the one law, which lead to a new system of Science altogether, that can be shaken by no difficulty, and is capable of explaining every phenomenon in nature.

In the first place :—All matter is eternal, and resolvable into atoms.

Atoms are invisible, indivisible, intangible, and indestructible.

They are separated into two great classes; viz.: mineral and vegetable atoms, or as they are at present called; oxygen and hydrogen. There are many different kinds of mineral and vegetable atoms, the former producing different minerals, and the latter uniting with the mineral producing different kinds of vegetation; *still there are only two classes of them.*

All atoms are Male and Female.

We find that all animals and vegetables are male and female, and as all *animate* matter is kept alive by eating or absorbing so-called *inanimate* matter—for the theory which divides atoms or matter into animate and inanimate, is untenable as we shew farther on—is it unreasonable to suppose that each inert atom is also either male or female?

These atoms also have in our world peculiar inherent properties belonging to each individually. For instance, the mineral atom is the Male, and its properties are, that it is naturally cold; that it has the blue and white cold colours; and that it is acid and combustible.

The vegetable atom is the Female; and its properties are, that it is naturally warm; that its colours are yellow, red, and the warm colours; and that it is incombustible.

Many Scientific men have endeavored to reduce matter to a simple form, but without success—Professor Tyndall in his lecture on "Matter and Force" at Dundee in 1867, said :—

"The matter of the world may be classified under two distinct heads, that of dead and living atoms. All atoms were once alive, but having exhausted their force in attracting other atoms in forming granite, limestone and metals, they are dead and cannot live again." This is manifestly incorrect, for by a simple experiment we can show that if these granites or metals were pulverized, mixed with vegetable atoms, and seeds were planted in them, and well watered, the atoms would show themselves alive by dissolving and aiding in producing a plant and probably a flower. Or they could be so mixed with other substances, that they would assist in blasting the hardest iron stone. *There is no such thing as a dead atom.* All atoms are alive or have inherent life properties, but they must occupy certain *positions* and *conditions* in order to show their vitality.

Prof. W. A. Norton in the American Journal of Science for 1872 endeavors to show that there may be only one kind of matter, and one form of force governing it. Matter is of three varieties, he says:—"Ordinary or gross matter, directly recognized by our senses; universal or luminiferous ether, filling all space; and electric ether, associated with all bodies of ordinary matter." He thinks however that they are all formed from luminiferous ether. The force which governs this matter is a law of *repulsion*. As the fact of an attractive power cannot however be denied, he seems to imply it is caused by the repulsive power of the atoms being so feeble that they attract each other! Rather a Quixotic mode of reasoning.

It would baffle him to explain how any phenomenon of nature is caused, and how any product of the earth grows, by the operation of the law of repulsion alone.

In the "Vestiges of Creation" we are told that matter was

originally a universal "Fire Mist," which gradually cooled off into suns and planets. How fire mist could cool when there was nothing to cool it, and how one substance could change without coming in contact with another, is inconceivable to us.

In the same book we find that there are fifty-five simple elements composing the earth.

But in an article on the "Materials of the Universe" in Fraser's Magazine 1869, we find there are sixty-two elements, of which forty-nine are metals, eight are substances with an individual character of their own, and five are gases. They can however be resolvable into the two classes of mineral and vegetable. The author of the above mentioned article says:— "We cannot affirm it to be matter of demonstration that none of these may be some day found reducible to a more simple form. We cannot pronounce with mathematical confidence, that no unexpected and startling discovery may yet effect at least a partial change iu some of these positions. But we may safely affirm, that the probability of any general revolution is infinitesimally small."

It remains for our readers to judge whether it has not now become a certainty.

We find the nearest approach to our views on the composition of matter, in Park's Chemical Catechism; where, after showing that plants may be grown in sand, litharge, and even in common lead shot, merely by moistening them with water, it concludes:—" Oxygen and Hydrogen, with the assistance of solar light, appear to be the only elementary substances employed in the constitution of the whole universe, and nature in her simple process, works the most infinitely diversified effects by the slightest modifications in the means she employs."

In conclusion there is a remarkable analogy between the component parts of matter, and the component parts of language.

Our alphabet is composed of a number of letters, but they are all divisible into the two great classes of consonants and vowels. By them we express ourselves in simple language that children might understand, while the inspired poet by the same means can give utterance to the grandest thoughts our literature contains. Although also the vowels and consonants remain the same, yet there is no lack to the multiplicity of new names that may be coined, nor a dearth of grand ideas, or of sonorous eloquence.

So in our language of nature, although it is composed of an infinite variety of simple elements, yet they are all divisible into two great classes.

By a simple union they form water and weeds, and worms and insects; while by a more intricate process they produce the beautiful bird, the pretty flower, and the lovely woman.

Moreover, as we have said by means of the same letters, new words are being coined, and new ideas expressed every day; so by the commingling of the same elemental atoms in nature, new plants are growing, new flowers are blooming, new colours are imparted, and new animals are created every day the world exists.

CHAPTER II.

MATTER AND ITS FORCE.

Atomagnetism.—What Prof. Huxley would like to know.—Matter and Motion.—Every Atom a Magnet.—Law of Atoms.—Like attracts Like.—Unlike poles Attract.—Atomagnetism the law of attraction and repulsion.—Examples.—Experiments with Filings.—How Atoms combine their Polarity.—Herbert Spencer's Philosophy.—His Foundation Loosened.

HAVING in the preceding chapter shown what the two great classes of matter are, we now proceed to explain the force which governs them in all their movements and products, and to which we have given the name of *Atomagnetism*.

It is this law which Prof. Huxley, as shewn in the following quotation, has been striving after without success. In an article on Bishop Berkeley's works in Macmillan's Magazine 1871, he says:—" There is a passage in the preface to the first edition of the Principia, which shews that Newton was penetrated as completely as Descartes, with the belief that all the phenomena of nature are expressible in terms of matter and motion." It is this " Many circumstances lead me to suspect, that all those phenomena may depend upon certain forces, in virtue of which, the particles of bodies, by causes not yet known, are either mutually impelled against one another and cohere into regular figures, or repel or recede from one another;

which forces being unknown, philosophers have as yet explored nature in vain."

That they may be unknown no longer, we will at once proceed to explain them.

A great deal has been said lately about matter and motion, but the latter is as yet only known to the most daring speculators as a senseless dance of the atoms; a dance so omnipresent and never ending, that they imagine it never had a beginning, and is as unlikely ever to have an end. On shifting sands like this the accepted theory of physics is based, forgetting that it is quite contrary to the working of any law of nature with which we are acquainted.

We have said all matter is formed of atoms. *But every atom is also we say a magnet having polarity.* That is, each atom has two poles similar to a compass needle, which may be designated north and south.

The law of atoms is observed to be, that like attracts like; but by the law of magnets, it is seen that like poles repel, while unlike poles attract. ATOMAGNETISM THEN IS THE COMBINATION OF THESE NATURAL LAWS, RESULTING IN A UNIVERSAL LAW OF ATTRACTION OF LIKE ATOMS AND REPULSION OF LIKE POLES. In order to make this principle clear, we will give a few examples.

A tree, for instance, in growing does not attract sand and clay and metal, and form a branch of clay or another of metal, but it attracts only material similar to itself. If a smith in casting a bar of iron tried to combine clay or coal ashes with it, he could not succeed, or else there would be no strength in the bar. Again we cannot attract the poles of a magnetic compass with a wooden stick, or a piece of coal, or indeed anything not in the nature of iron or steel.

This is the attraction of like atoms.

Now suppose we have two bars of magnetized iron, and some iron filings. If we take the ends pointing north, or the ends which attract the north point of a compass needle, and dip them into the filings, these particles will stick to the bars, and bristle out on both ends like hairs. Suppose we then bring the same ends together, the filings seem possessed of life and fall back from each other, thus showing the repulsive power of similar poles. But if we place filings on the north end of one, and on the south end of the other, the particles stretch out from each bar and cling to one another. This is a simple experiment, but very important and suggestive, and it may be performed by any one.

While thus all atoms are magnets having polarity, when a number of them coalesce or join together, each individual atom merges its polarity into the force or polarity of the whole.

For instance, a small piece of iron has a north and south polarity, because it attracts the points of a compass needle; but suppose we take some hundreds of these small pieces, and form them into a large bar of iron, we do not find some hundreds of north and south poles in it, but only one north and one south pole. The force which was in each small piece is concentrated in the whole, so that the attractive and repulsive power of the bar is stronger by every piece that is added to it.

While then it may be believed, by our readers generally, that some atoms are magnets, as iron for instance, it will be hard to convince many that all atoms are magnets—for instance, the atoms of the paper we write on, or the bread we eat; but it shall be our endeavour to prove them to be so in succeeding chapters. It will be hard also to convince the world that besides being magnets, all atoms are governed by the universal law which we have advanced, but not till the last chapter is finished, will we consider that we have proved our assertions, for should one effect or phenomenon of nature arise which cannot be explained by it, then we throw the whole theory aside as utterly worthless. We conclude this one by stating, that as Herbert Spencer's Philosophy is based on the fact, that all matter obeys a simple and universal law of attraction, and as we have now shown there is also a law of repulsion, his foundation is defective, and the greater part of his theories must consequently be worthless.

CHAPTER III.

MINERAL LIFE.

Minerals not dead.—Mineral life a low form of vegetable and animal Life.—Iron filings have life.—Compass needle has life.—Philosopher's tree.—Coral.—Candy.—Mineral life.—Atomagnetism.—Atoms of lead, sugar, and coral, Magnets—Greater always influences the less.—Explanations of Philosopher's Tree.—Cause of beautiful forms in snow flakes.

The title of this chapter may seem curious and incomprehensible to many, but we do not think it difficult of explanation or incapable of being understood.

Minerals have been looked upon as dead matter, but as we have said before, every atom in this universe has life, but it must be in a certain position and condition to show it.

Mineral life is indeed the lowest form of life. It is not so complicated as either vegetable or animal life, yet it is caused by the same law. The experiment of the iron filings shows that they are possessed of life. A low form of life it is indeed, yet one that infallibly directs them when to attract, and when to repel. The compass needle has similar life, and man although possessed of the highest form of life on this earth, has to depend upon its guidance, and to stake his life upon its warnings.

But there are other forms of mineral life. We all know or have seen the philosopher's tree. A piece of zinc is suspended in a solution of sugar of lead, and in a short time the atoms of lead will be seen deposited on the zinc; which will then appear to shoot out branches and leaves in the form of a tree. If these particles of metal are not possessed of a certain principle of life activity, what leads them to arrange themselves in such a beautiful manner?

We have another fine example of growth in a similar manner in the coral—erroneously supposed to be formed by insects, but shown by us in another chapter to be the result of natural growth from water, saturated with suitable material.

All confectioners too in making crystalized candy, see the operation of the same life in the formation of it, although they do not understand it otherwise than as the process of crystalization.

Mineral life is also exhibited in every snow flake that falls. The beautiful forms which they exhibit, all springing from a centre—being caused by the same law.

Mineral life is only one form of atomagnetism; and it shows that all these various substances, the particles of lead, the coral, the sugar, and the snow, are all magnets. For instance, in the philosopher's tree, the piece of zinc first obeys the atomic law, and draws all the small atoms of lead in the solution to itself—as the greater body always influences the less. But being a magnet also, it has polarity, and as particle after particle is added to it, its force is continually becoming stronger, so that it shoots out branches and leaves as though possessed of vegetable life.

Thus, in conclusion, while minerals have life, it is not a form of it fitted to produce seed, but in other respects it is like the vegetable and animal life, inasmuch as it reproduces its kind from suitable material.

CHAPTER IV.

VEGETABLE LIFE.

Origin of Life.—Spontaneous generation.—Sir William Thomson on seed bearing Meteors.—Cornhill Magazine.—Atomagnetism.—No seed required.—Railway Cuttings.—Clover.—How a plant grows without a seed.—Scripture proof for it.—Seeds rot.—Hardwood and soft wood Forests.—Darwin's "Origin of Species" overthrown.—Thousands of plants in the first creation.—New plants with every change of soil and climate.—Present theory of plant life.—How a cell develops.—Absurdity of plants breathing.—Why roots and branches spread.—Experiments to prove the reason.—Why a tree does not grow in winter.

For a number of years of late, there has been considerable discussion on the origin of life upon the earth. Vegetable life has been accepted as its lowest form, and numberless theories have been advanced, some in favor of spontaneous generation, but most against it. The latest and most extraordinary theory is that advanced by Sir Wm. Thomson.

In a lecture before the British Association, he suggested that Meteors were *seed bearing*. For instance, two worlds or more collided in space and burst; and the fragments bearing moss seeds found their way to our earth; and thus generated life on it. But supposing this correct, his theory would not account for the origin of life in the universe; there must have been a higher power to originate the seed bearing meteors.

A writer on this subject in the Cornhill Magazine for 1872 very aptly says:—"By this theory, nature must destroy two worlds in order to plant a few moss seeds in a new one."

Following up the law of Atomagnetism, we say, that the growth of any plant is governed by the same law which governs the philosopher's tree; and that, practically, a seed—in the first instance—is no more required in the one case than in the other.

We are told that no vegetation can be produced without a seed; but how is it that when a railway cutting is made in new earth, which has never before been disturbed, a crop of clover should immediately grow up; and that it, in its turn, gives way to other vegetation, as the soil changes year by year.

Besides, every seed does not become a plant; for they often rot by being placed in unfavorable positions for growing; and thus they return to the dust from which they sprung. If, then, seed can return to dust, is it not possible for dust to perform the functions of seed?

Moreover, a plant, while growing, forms a seed in itself from the soil in which it is placed. What, then, should prevent seeds or plants forming in the soil itself, under favorable circumstances and suitable conditions? Let us detail the process by which this could be accomplished.

In the soil we have mineral and vegetable atoms. All atoms are magnets. Several *like* vegetable atoms are attracted together, along with a sufficiency of mineral atoms to generate a combined action. Rain, composed also of both classes— oxygen and hydrogen—descends, and increases the force. More atoms are attracted; and with the addition of every atom, more action or life is given to the whole. In due time, enough atoms have come together to enable the force to show itself. Accordingly, a root is thrown out below, and a leaf is pushed out above. The further progress of the plant then is simple. The root attracts atoms in every direction from its battery—the decomposing soil—and sends them as sap to the axis of the plant, which is the life centre. From there they are sent through the whole plant by the nature that governs them; either to the branches to develope leaves, blossoms, fruit, or seed in succession, or to multiply and extend the roots.

It may be asked, Why should a seed be formed in the plant at all, seeing there is no absolute need for it? This question would lead us still further, and we would ask why there should be any fruit, or blossoms, or any law ordering plants yielding seed to be formed from the soil, and who ordered the same? Scientific men in general are not noted in their theories for paying much regard to the statements of Scripture, nor are they willing to look there for authority in discussing them. But in proof of the correctness of our views, we can point in many instances to Scripture for our proof. There are in the Bible many mysteries which man does not understand, but when he shall thereby be fully informed, little else will be left for him to learn.

In confirmation of, or agreement with, the correctness of our theory of vegetable life, we quote the following verse from Genesis:—"And GOD said, *Let the Earth* bring forth grass, the herb yielding seed, and the fruit trees yielding fruit after his kind *whose seed is in itself,* upon the earth; and it was so."

The thoughtful reader will observe that it is not written:— " As God said, let the seed of the grass be planted, that grass may be brought forth by the earth;"—but he ordained: " *Let the earth bring forth grass."* In other words, God in creating the earth, had placed within the atoms of which it is composed, that principle of life giving force, which is part of His universal law; and by which the earth, spontaneously, brings forth the " grass, the herb yielding seed, and the fruit trees yielding fruit after his kind."

Surely He who gave the atoms their atomagnetic power and properties, is able to give them any other.

Again, plants die out, and seeds do not always grow to plants.

For instance, a forest of hardwood is burned down; and, although the roots may be uninjured, the hardwood trees do not renew themselves, but a forest of softwood grows up in their place. What is the cause of this? The nature of the soil has been changed by the fire, and the new soil, as a natural consequence, not being adapted for the hardwood, developes a plant bearing seed suitable to itself.

Thus is Darwin's theory of the "Origin of Species" overthrown in a moment. He asserts all plants are descended from four or five progenitors, "or probably only one." But if a plant will grow only in a soil, and in a climate suitable to it, there must have been some thousands of species at first. For Darwin surely does not mean us to believe, there was only one kind of soil, and one climatic condition in the beginning.

We do not, however, believe with his opponents that no new species have sprung up, for with every change of soil there must necessarily be a change of plant growth. Every day, therefore, old plants are dying out and new ones are being created. Experienced gardeners, while they can produce a new variety of almost any plant or fruit grown, might, doubtless, if they tried, produce entirely new plants also, by chemically changing the soil.

Having thus given our view of vegetable life, it may not be uninteresting, to see how it tallies with that given by botanists.

Although botanical students know the structure of every well known species of plant, and examine the minutest parts of them with a microscope, besides giving them names, such as cells, granules, raphides, &c., yet, strangely enough, they have not attempted to discover the principle of *force* which makes them grow. They say a cell contains the first germ of life, that it grows by dividing itself into four other cells, and these four into sixteen, and so on. *But what impels the cell to divide itself? there is the mystery!* They say plants have two principal parts, an ascending and a descending axis. (See how agreeably this comes in with our theory of the poles.) The one axis produces leaves and branches, whereas the other only forms roots. In the leaves they find flattened cells with mouths, and consequently they say plants inhale carbon and ammonia. In the roots they find spongeols, and these they say draw up water and potash only.

If we ask, Why the roots spread out in all directions, it is to search for food; and why the leaves and branches spread out in the atmosphere, it is to give them room to breathe. If we ask what the power is that causes them to do so, they cannot explain; unless that it is a wonderful law of providence for preserving the life of the plant.

We will now refer to the fallacy of those botanical statements.

That plants breathe through their leaves, is just as absurd as to say a man breathes through the pores of his skin. They say this is proved by coating a tree with paint and it dies. But if we coated a man with paint, he too would die. The reason of this is, the exhalations from the body through the pores are stopped, and sent back, contrary to the healthy course of nature, and the man dies. So it is with the tree. It is giving out exhalations continually by its flattened cells, and should they be checked, the tree is choked, and its life action ceases.

Moreover, if we placed a man under a glass case with plenty of food and water, and closed every aperture, he would speedily die for want of fresh air. But, enclose a plant under the same, and it thrives and grows well; thus showing that plants neither breathe, nor inhale carbon, or anything else, from the atmosphere. The flattened cells answer a similar purpose to our pores, and are merely the channel of exit, for the exhalations from the plant.

We will now explain the force which naturally spreads the roots and branches.

Suppose we again take the magnetized bar of steel, and the filings. If we dip both ends into the filings, we have a perfect representation of a tree in winter, with the roots at the one end, and the bare branches at the other. No two roots in a tree are observed to go together, nor any two branches. Even the smallest twigs are observed to shoot out in places where no others are growing. So it is with the filings. Each hair of steel dust is as distinct from another as needles would be. The reason of this is—as already explained,—that similar poles repel each other. Roots and branches of the same tree, therefore, repel each other, because they are of the same poles. A plant or tree consequently must have polarity, and this would show us, if we did not know from other sources, that *a tree is by nature a magnet.*

It may be asked why a tree in our latitude does not grow in winter? This is a question that cannot be answered by a botanist. For, if plants feed on carbon,—which they say is given out by animals—they ought to grow in winter as well as in summer, for animals are breathing at all times of the year. How is it also that they obtain their carbon in a hot house in winter, from whence all exhalations except the gardener's own breath are excluded? Just for the same reason that we find trees in tropical climates, growing and bearing flowers and fruit all the year round. They need warmth, and similar exhalations to their own to grow in. A cold, mineral, frosty air, checks the exhalations, and kills the plant. While a warm, vegetable, moist, (or mixed) atmosphere, draws it out and increases its growth.

CHAPTER V.

ORIGIN OF ANIMAL LIFE.

Man afraid to inquire into the origin of life.—Milk and cheese.—Dumas and Agassiz on seeds and eggs.—A cow the mother of maggots.—Insects spontaneously produced.—How animals are produced without an egg.—Excess of vegetable matter forms animals.—Process of creation.—Darwin.—All animals produced not from one but from many.—Agassiz on men and monkeys.—One animal may produce a different animal.—Animals parasites.—Argument against spontaneous generation.—Germ theory.—Pasteur.—Child.—Lamarck.—Canned meats.—Why ice and salt preserve meats.—The formation of germs.—Tyndall on respirators.—Spontaneous fish.—Agassiz on special creation.—Origin of lowest organisms.—Mr. Charlton Bastian.

HAVING shown that mineral and vegetable life are caused by the simple law of atomagnetism, it will be said we are surely not giving animal life so humble an origin. Human life has ever been considered such an awfully mysterious thing, that men come to look at it as something forbidden to be talked about. "God breathed into man's nostrils the breath of life" has been sounded faintly through all the ages, into intelligent men's ears, and the meaning conveyed thereby was so inscrutable, so sublime, and so far above mundane thoughts, that few ever dared to inquire into it.

No wonder that men should have such a dread of it, when everything connected with life is so unfathomable.

A great poet full of lofty ideas, whose works are known by educated men all over the world, by an accident falls and breaks a cord in his body, and at once he is as inanimate, as senseless, and as incapable of anything for good, as the stones he lies on. That head once brim full of knowledge and overflowing with song, is now vacant and silent as the tomb, and the key to those chambers of learning the world wondered at, is lost forever. But the mode in which life comes into the world, astonishes us as much, if not more, than the way it disappears.

We buy a piece of cheese, and in a few days it is teeming with living creatures; or we lay aside milk for the same term, and the microscope shows us thousands of living organisms. Who were the fathers and mothers of these creatures?

M. Dumas in his first Faraday lecture, denied that the chemist with all his endeavours had ever imitated life itself, or would ever be able to produce a living being. "There must be a living seed for a living plant, and a living egg to produce a living animal. These were far above human power, and within the power of God alone." Agassiz also says:—"All living beings are born of eggs, and developed from eggs." Did the cow then furnish the eggs from which these organisms grew in the cheese, and the milk?

In the "Vestiges of Creation" we find that a Mr. Crosse, unexpectedly, produced insects while conducting some chemical experiments with silicate of potash, and since that time, numberless experiments have been tried with different materials, ending in similar results. What inference therefore must we draw, but that life is regulated according to some law which God gave to the component materials of this earth, and that when the full conditions required for producing life are fulfilled, the so called creation of minute animals, is as natural as the growth of weeds from the ground in spring.

Let us then explain how animal life may be produced from matter without an egg.

In all life and growth water is necessary. Water is a combination of the two great classes of matter, vegetable and mineral, or as it is commonly called oxygen and hydrogen.

In mineral life and growth, an excess of mineral atoms in watery solution is requisite.

In vegetable life and growth, both mineral and vegetable atoms in watery solution are required.

And in animal life, an excess of vegetable atoms in watery solution is essential.

It may be asked why an excess of vegetable atoms should form animals, and not vegetables? Because in vegetable life, a certain quantity of mineral atoms, and a certain position and condition are required; and should these not be obtained, an animal results. For instance, if we moisten some hay seeds and expose them on a warm floor for a few days, animal life will teem all over them. Why is this? Because the requisite amount of mineral matter could not be obtained. If they had been placed in the ground, grass would have grown. Thus the difference in the production of animal and vegetable life, is chiefly one of *condition* and *position* of matter.

Let us explain how these minute animals are formed. Every atom is a magnet. The water and heat in the atoms of hay, initiates an action, which leads them to form an attraction of like to like. A number of these atoms being brought together, an action commences in the centre as a stomach, a mouth opens to take in other atoms; and thus an animal is the result. Low animal life is thus only a higher form and development of vegetable life, the animal as a general rule, being migratory, and the vegetable stationary.

Wherein, then, consists the life of this animal? Merely in the process of the stomach, receiving the food and dissolving it, and the innate force of the animal as a magnet, sending the necessary material from the food as blood, to its extremities to repair and increase its body.

All animal life including that of man, consists therefore, first, in the dissolving of food into atoms, and assimilating them to the material of the body; and secondly, in the magnetism of all the atoms of the body acting as one magnet, forcing these atoms from the centre to the extremities of the body.

The law which governs life then is ATOMAGNETISM. In corroboration of our statements we have that of Prof. Leo. H. Grindon, who in his valuable work on "Life" says:—"All life, whether physical, physiological, or spiritual, is a state of marriage or the union of two complementary forces acting and reacting."

Such being our theory it may be asked, do we then say with Darwin, that all animals have been produced from one minute primordial form, and have so progressed up to man? By no means. Just as matter itself is composed of a variety of elements—just as all minerals are not, and cannot be produced from one sort of mineral matter; and all vegetables from one species of vegetable matter; so animals are governed by the same law, and have not been produced from one animal, but from many.

Agassiz is of the same opinion, and even carries it further when he says, in his lecture on "Men and Monkeys," 1867:—"If it is an error" (as he proves it is) " to consider man as derived from monkeys, we must admit that men are not derived from a common stock, because the differences which exist among men, are of the same kind, and quite as striking, as the differences which exist between monkeys and the lower animals." Also : —" The doctrine which I support, is that it is not only the few which were started in the beginning by a creation act, but the many."

We do not say however that one animal may not produce some other, just as some plants produce others. Wheat, for instance, if sown one year and cropt during the summer and autumn, will produce a crop of rye the following season.

All plants and animals may be classed as parasites. Not only has each variety of mineral its peculiar soil, and each soil a plant peculiar to itself which may be considered its parasite ; but in almost every instance, the plant itself has other smaller plants or animals parasitical to it. Each animal also, while a parasite to certain soils, has other animals parasitical to itself, and so on indefinitely.

Raspberries produce an insect peculiarly their own, so do strawberries, apples, figs, and in fact every kind of fruit. Yet these same fruits, when preserved, will produce a *different* kind of insect from that which it produces in its natural state. Thus showing that the nature of the animal, is dependent on the matter it springs from, and feeds on.

All animals including man are parasites of the earth, and each is peculiarly fitted for the climate in which it is a native.

The camel, the elephant, the lion, the moose, the polar bear, the seal ; in fact each particular animal, enjoys its life to the utmost in the place where it is found ; and speaking generally, when animals are transported to other countries, they live a precarious existence, and require to be carefully tended—a study being necessarily made of their natural appetites and habits, in order to preserve life.

We find also that many animals die, or are exterminated from a country. The wolf has gone out of Britain, and geologists tell us that the elephant, the woolly rhinoceros, and the cave bear, had once a home there in ages gone by.

What inference then must we draw from these facts? Nothing less than that these animals were produced on those soils, and in those countries in which their remains are found, and that the condition on which their lives depended being changed, they had to die, and became extinct.

While thus showing the truth of spontaneous generation, let us review the arguments which have been adduced for and against it.

A number of years ago, the French Academy offered a prize for the best essay on the subject, and it was awarded to M. Pasteur, a celebrated chemist, an opponent of spontaneous generation. Dr. Child, an eminent English physician, however, tried the same experiments from which Pasteur derived his arguments, and with a more powerful microscope, found life where Pasteur could not see it.

In order to explain how fungus mould on cheese, and on boots, mildew on cotton, the hop blight, and the vine disease are caused, Pasteur says, "the air is filled with living *invisible* germs, which alight in suitable places, and commence to grow immediately." As this germ theory is the main argument adduced by the opponents of spontaneous generation, let us see what it really means.

Life we are told is originated by living invisible germs. What is the difference between a germ and an atom? We do not see that there is any difference, but many scientific men seem to say there is.

We explain what atoms are, where they come from, and how they cause life; but where do these germs come from? *They must have been born in some way.* Pasteur does not know, and the only man of science who advances an opinion regarding their origin, is Lamarck. He says that the germs or rudiments of life, which he calls monads, are continually coming into the world, and that there are different kinds of monads, for each primary division of the animal and vegetable world.

"This hypothesis," as Sir Charles Lyell correctly says, "is wholly unsupported by any modern experiments or observations, and affords us no aid whatever, in speculating on the commencement of vital phenomena on the earth."

Lamarck's theory, to be of any service, ought to have explained where the monads came from.

If invisible atoms, and invisible germs are not the same, then the germs must be formed from atoms, by the law of atomagnetism; that law or cause of life, which men have been seeking for so long, and could not find. We can prove that germs are not exclusively in the air, and that atomagnetism causes germs—if such things do exist—for Pasteur says, in speaking of the length of time during which preserved meats and vegetables may be kept in cans:—"this result is solely obtained by the exclusion of the germs of corruption and decay, which prey upon all perishable substances, with more or less rapidity."

But if we expose one of these cans to the heat of the sun,

in a window, for several days, we find that *germs have gained admission to it*, and animals innumerable, have been hatched from "nothing." Although the can was air-proof, the heat of the sun, acting on the meat and the water, or the vegetable and mineral matter inside, formed a power which immediately set the atoms moving. Like began to draw to like, germs—as they say—were formed, *magnetism* set in, and in a few days, the so-called can of dead animal or vegetable matter, was a mass of moving and living animals.

Moreover, this preserved meat, if taken out of the can and placed on ice, will keep for months, although exposed to the atmosphere, thus showing conclusively, *that germs are not floating about in the air alone, but are formed from matter under certain conditions of temperature &c.*

It may be asked, why ice and salt preserve meat so well. Merely because they are mineral substances; and as we said before, animals are only formed and supported, from an excess of vegetable matter, kept at a certain high temperature.

Professor Tyndall believes in the floating germ theory, and in an article on "Dust and Disease" in Fraser's Magazine, he shows the air of London to be so bad, that he recommends every one to wear a woolen respirator over the mouth. This is unnecessary, for the nose is so constructed that a germ would have great difficulty in reaching the lungs by it. The only precaution required is to keep the mouth shut.

Another difficulty with the germ theory is, that while its upholders believe it to develop into fungus, worms, and insects; they cannot understand how it develops into fishes. For instance, the writer of an article on "Spontaneous Generation" in Blackwood's Magazine (1861) says:—There are numberless cases in which we are baffled, in the attempt to explain how animals could possibly find their way to the places where they are discovered, but spontaneous generation is not an acceptable solution of the difficulty. No one supposed that the fish which Macartney found in a pond, in the middle of an island far away from any continent, and which seemed to have been thrust up from the bed of the ocean by a volcano, were produced spontaneously; *yet how they got there is inexplicable.*" Why should they not be produced spontaneously? Small animals would develope out of the vegetable matter which grew in the water. These would enlarge and alter, as the water gradually changed from the different substances draining into it; and at last, a full grown fish would be developed, suitable for living in the pond, and capable of propagating its species too. In this view there is nothing inexplicable about it. Spontaneous generation, is merely the result, and *continuation*, of the law and order of Creation at the beginning.

In opposition to the theory of spontaneous generation, or Evolution, its sister theory, Agassiz maintains the doctrine of *special creation*. But we perceive no material difference between special creation and Evolution, as we have explained it. For instance, would it be more difficult for the Creator to order at the beginning for all time, that just as certain materials were mixed together under certain conditions —such as warmth, moisture, and air—an animal would result, of a character and kind suitable to the material it was born in; than to interpose in a special manner, whenever any chemist should be chancing to mingle *creative* substances, or wherever a new kind of preserve was storing, or some fruit was decomposing?

It would be limiting the power of the Creator to say so.

It is evident, however, that while matter has been endowed with an inherent creative force, both in plant and animal life, a certain limit has been affixed to that power. No being higher than man may be formed; no monstrosity may be perpetuated; nor any animal created, but what may be controlled or governed by man.

If a simple evolution of nature originated all things, we would have expected that in the course of six thousand years, all these above results would have happened. We should, for instance, have found some men able to fly. If, as Darwin says, animals develope into something higher by merely wishing to advance; then certainly it is not for want of wishing, that man is not gifted with the power of flight; for, from the days of the mythological Icarus,—who soared so high that the sun melted the wax on his wings, and he fell into the sea—there have not been wanting men to experiment with wings on their shoulders, but the appendages refused to become incorporated with their bodies, and all their endeavours have only ended in failure, if not disaster.

The nearest approach to our explanation on the origin of life, is in a pamphlet by Mr. H. Charlton Bastian on "The mode of origin of lowest organisms;" which is a reply to Pasteur, Tyndall and Huxley. After detailing his experiments, he sums up by saying:—"It would thus appear, that specks of living matter may be born in suitable fluids, just as specks of crystalline matter may arise in other fluids. Both processes are really alike inexplicable. Both products are similarly the result of inscrutable natural laws; and what seems, inherent molecular affinities. Living matter developes in organisms of different kinds, while crystalline matter grows into crystals of divers shapes."

The only thing he appears to lack, is a knowledge of the *inherent molecular affinities*, and that, it will be evident, is supplied by nothing else than Atomagnetism.

It will be seen therefore, that Atomagnetism explains all difficulties connected with the subject of life, and that a knowledge of it, is all that is necessary for every one to see the mode of, and to believe in the process of, spontaneous generation. It accounts for Lamarck's monads, explains to Mr. Bastian how animals are born in suitable fluids, and finally annihilates the germ theory.

Life, awful, mysterious life, is then merely that atomic attraction, and repulsion, of matter, which we see everywhere combined with, or similar to, that magnetism which is exhibited so abundantly in steel. This simple law (for the one cannot work without the other) is so universal, that in future chapters we will show, there is nothing so minute that it does not affect, and nothing so powerful that it does not control. The tiniest shell fish that creeps on the shore, and the noblest animal that walks, are governed by the same law. All movements whatever are but parts of the one great machine the globe, our earth itself being, so to speak, but a wheel in the grand clockwork of the universe.

CHAPTER VI.

APPETITE, OR INCIPIENT MIND.

Darwin thinks development of Mind a hopeless inquiry.—We explain it.—Appetite the lowest form of Mind in Animals.—Spontaneous Insects eating immediately.—What is Appetite?—The Atomic Law of Like to Like.—Mind and Life, Properties of Matter.—Vegetable Appetite.—A Seal's Appetite.—A Calf's Appetite.—Why it does not eat bricks and stones.—A Baby's Appetite.—Appetite for Tomatoes.—Superiority of a Brute's Appetite over Man's.

DARWIN in his "Descent of Man" says:—In what manner the mental powers were first developed in the lower organisms is as hopeless an inquiry as how life first originated. These are problems for the distant future, if they are ever to be solved by man."

But if we have shown how, by an atomic law, the greater mystery of life originates, surely an explanation of the lesser mystery of mind ought to be comparatively easy.

All animals, the instant they are endowed with life and freedom, are given also a mind suitable to their condition, in the form of an *appetite*; which is the lowest phase of mind in animals. For instance, the minute insects which Mr. Crosse brought out of silicate of potash, immediately began to feed on the matter they were produced from. What led them to do so? What told them that silicate of potash was good for them to eat? Their instinctive appetite, of course. What then is the natural appetite? Merely the atomic law of like attracting like, the same law which draws similar materials to minerals and vegetables, to assist in their formation. The matter of the animal, having an affinity for material like that it is formed of, is drawn towards it, and feeds on it. The animal then is, as it were, merely a living magnet, and its appetite or mind, merely a *property* of the substance of which it is composed; as magnetism is a property of the iron magnet.

Incipient mind, or appetite, and life, are thus alike manifested as properties of matter, in different conditions only.

Let us now see how this same law of appetite governs all animals, from Mr. Crosse's insect up to man; and not only animals, but vegetables too.

All kinds of vegetation will not grow on the one soil, for each plant has an appetite, and this can only be supplied by a situation which contains materials similar to itself. If planted in a different soil it dies. The same law holds with animals. We have seen seals transferred from the clear, cold, salt water of the North Atlantic Ocean, where they had abundance of fresh fish for food, and placed in a pond of muddy fresh water, in the belief that they would enjoy life in company with frogs, eels, and muskrats. They died of course.

When a calf is born, it is immediately attracted to the cow for a supply of food, because its natural appetite tells it, its mother can furnish suitable material; and shortly afterwards it proceeds to nibble the grass, and to drink water. The material of the calf's body being produced from grass, saturated with water, it has an attraction to a similar vegetable product growing in the field, and to water, and thus it partakes of both.

For this reason it does not proceed to chew bricks, or stones, or rubbish, as we might expect such a young and inexperienced calf to do; nor will it eat a different vegetable from that fed on by its mother; so powerful is the atomic law governing its life and instincts.

A baby likewise is drawn to its mother, and feeds on her milk, or on substance similar to the food which it obtains from her. Its appetite as it grows to manhood also, is still governed by the materials composing its body, and it generally dislikes what it never tasted before.

For instance, a Briton going to America, and trying to eat tomatoes, Indian corn, or bananas, generally dislikes them, till by persistence in tasting them, the material becomes incorporated with the other material of his body, and he finally acquires an appetite for them.

There is one grand distinction, however, between the appetite of the beast and of man. The beast, uncontaminated with man, will never eat anything which is injurious to it, as it is perfect and never fails; while man's appetite, guided by his reasoning powers, is uncertain, and not to be depended on. Why this is so, is explained in the next chapter.

CHAPTER VII.

INSTINCT, OR ANIMAL MIND.

Instinct a higher phase of Mind.—Frank Buckland.—Why a Chicken knew a Gentleman was not its Mother.—Sparrows require no Teaching. — Foreknowledge of Bees and Beavers.—Important Fact.—Animal's Mind Perfect—Never Progresses.—Man always Progressing.—Difference between Man and Beast.—Man two Minds.—Animals one.—Animals no Soul.—Mind returns to Earth.—Their Mind all Nature.—Animals Perfect on separation from the Parent.—Answer to Frank Buckland's questioner.

WHILE appetite is instinct, and should hardly be separated from it, yet we wish to allude more particularly to the higher phase of animal mind, as shown in the skill of the bees, ants, and spiders; the foreknowledge of the beaver, the cunning of the fox, and the scent of the hound.

A curious paragraph went the round of the newspapers some time ago, about a gentleman, who told Mr. Frank Buckland, the naturalist, that on watching a chicken coming out of its shell, it ran away from him as soon as it was free; the gentleman wished to know how the chicken knew he was not its mother. The question was a puzzler for the naturalist, so he contented himself with merely turning the whole thing into a joke. But there is more than a joke in the query, and it involves other questions about the instinctive habits of animals, which should have definite answers in this age of intelligence.

There is something very curious about the habits of what we call the inferior animals, which seems to make them gifted in many respects, far beyond what man apparently will ever attain to. Instinct in them often excels reason in man. We find the inferior animal, as it comes into existence to be perfect—so to speak—each according to its species and its instincts; as shown by each adapting itself to its proper food and conditions; whereas man is altogether helpless, and he only reaches a kind of perfection, by the long road of education.

The scent of the hound, is a gift which no man possesses; and the foresight displayed by the bee, in laying up a store of honey when the winter is to be long, and in killing the drones early if a wet summer is coming; seems to baffle human prescience entirely. The swallow also, in his mild winter residence, knows when it is time to migrate to his northern home, and the carrier pigeon, no matter where taken to, will find his way back to his original starting place: the foreknowledge of the beaver and muskrat, outstrips the telegraph, while the spider, excels man in the art of spinning, and drawing mathematical lines. Yet there lies the comparative limit of their skill; for, while the animal of to-day knows no more than his congener of creation, the man of to-day is continually progressing in knowledge and acquirements.

What then constitutes this immense difference between man and the beast, and how does the latter inherit it?

The grand difference lies in the fact, that the animal is possessed of one mind only, known as instinct, and that it is perfect in its kind; while man has two, an animal mind, similar to that of the beast, and a divine one, the soul, implanted by God.

The beasts have no soul or spiritual existence. Their mind being merely a *property* of earthly matter, and of earthly origin, returns to the dust from whence it came. They know not sin nor do wrong. Being guided by nature, all species of animals have their own instincts. They have appetites, but no desires. Their only object in life is to satisfy their wants, to preserve and defend themselves from danger, and to increase their kind. Thus it is they know what to eat, and what to avoid, while they are also provided with a suitable defence in danger. The fox has cunning, the flying fish wings, the porcupine bristles, and so on indefinitely.

But how is it they can know beforehand of the near approach of storms?

If their appetites, or the material composing their bodies, infallibly regulate them in their choice of natural food, and of the unfailing remedy to be taken when they are sick; what is more probable than that the gaseous adjuncts of their minds, being of similar material to the atmosphere, should inform them of every change about to take place in it. If so-called inanimate substances like mercury, salt, etc., should be so sensitive to atmospheric changes, how much more would an animate body feel them, *whose mind we might say is all nature itself?*

It may be asked, how is it that the mind of the young animal is so perfect? Because it is transmitted by the parent.

If a hen gives to her brood, wings, beaks, legs, toes, feathers, &c., similar to herself, and not similar to a duck; she must give them appetites, and a mind, also, similar to her own. If a chicken did not derive its mind from its mother, it would not necessarily act like a hen, but might follow a duck into the water, or try to quack, or do many other stupid things which would be considered unnatural.

We can now answer Frank Buckland's questioner by saying, that the chicken knew he was not its mother, firstly; because its appetite told it, there was a deficiency of proper food for it in the person of the inquisitive gentleman; and secondly, because its mother had a natural dread of such persons, and communicated that feeling, as a matter of course, to her offspring.

It may be asked, why are foxes cunning, geese stupid, and ants industrious? This can only be answered by asking, why is sugar sweet and salt acid? Each mineral has a quality of its own, gold, silver, iron, and granite, for instance. Each plant has a property of its own. So also each animal has a character of its own. Why they are so and should not be something different, is beyond our inquiry; as even why certain letters should represent a particular sound. They are only known to us as properties of the material forming each, and every one of them is developed in many, if not in all cases, by the circumstances which surround them.

CHAPTER VIII

MAN'S ANIMAL AND SPIRITUAL MIND.

Schelling and Hegel on Nature as "petrified intelligence."—Hope on "Origin and Prospects of Man"—Matter without properties.—Mind a property of matter.—No limit to the properties of Matter.—Brutes have one mind, man two minds.—Animal and Divine.—Agassiz on two minds.—Why his animal mind degenerated.—Man should distrust man.—Manner in which mind is formed.—From food.—Difference between animal mind and Divine.—Situation of the mind.—Of Memory.—Brain a picture gallery.—Difference between man's mind and the brutes.

SOME philosophers assert that mind is matter. Schelling and Hegel, for instance, say that surrounding things are "solidified mind," and nature is "petrified intelligence."

Hope, also, on the "Origin and Prospects of man," says:— "Can we say that God has not in matter itself laid the seeds of every faculty of mind, rather than that he has made the first principles of mind, entirely distinct from that of matter? Cannot the first cause of all we see and know, have fraught matter itself from its very beginning, with all the attributes necessary to develope into mind."

But we can fancy matter without any properties, where the world would be a perfect "chaos without form and void;" as it was in the beginning; and once this is admitted, mind cannot be matter, but only *a property of it.*

Although the matter of this earth has been endowed with atomagnetic properties, we have no right to say all other worlds are endowed with exactly the same force; for we would thus be limiting the power of the Creator. If, then, matter can be endowed with different kinds of properties, and we find it to be so, there is no limit that we can ascribe to the properties and combinations of matter.

If therefore man is endowed with an animal mind, possessing the properties of the matter of this earth, in character similar to the beast; who can say that he is not endowed with an additional divine mind, not included in our earth's material properties at all?

If we say that the beast's mind is as perfect as the properties of this earth can make it, and that the brutes worship no superior being; we must admit that man has a Divine mind, in addition to his earthly mind, because he believes in and worships a superior being.

In Agassiz's lecture on "Men and Monkeys" delivered in 1867, we have a partial confirmation of our theory. He says:—"Were we not made in the image of the Creator, did we not possess a spark of that divine spirit which is a godlike inheritance, why should we understand nature? Why is it that nature is not to us a sealed book? It is because we are akin to the world, *not only the physical and the animal world, but to the Creator himself*, that we can read the world, and understand that it comes from God." Agassiz, however, while thus obviously suggesting two different minds, yet did not grasp the idea in all its fullness, and consequently is weak and erroneous in many of his arguments.

That the *animal* mind when properly cultivated in man, is

not greatly inferior to the beasts, may be seen in those tribes lowest in the scale of civilization, who live and prey upon animals. The manner in which one Indian will track another through the forest, is almost equal to a dog's faculty of scent. The instinct which shows the former how to make a fire; to know the signs of the weather; which tells the natives of northern latitudes to eat no salt, and to drink no cold water in winter; while their more civilized neighbors bring on sickness and disease by overdosing themselves with both, all show that the animal mind only wants cultivation. He can also outwit the fox, and outstrip the swiftest antelope with cunning; besides he can bid the elephant do his will, and is also able to combat the lion with advantage.

How is it then since man at first rose above the brutes, that his animal mind proved to be of such little use to him; that as he progressed in knowledge, skill, science and art, he knew so little of the weather; that he distrusted his appetite; that he could not cure himself when sick, and died more by ignorance than disease? Surely man was not given a divine mind to acquire a superior knowledge of nature and art, and to forget what concerns his life, and the best way of preserving his health and existence on earth. Certainly not. His Divine mind, if consulted and properly controlled, was given to aid him in understanding all that the brutes know, and infinitely more; but man distrusted or misused it. He preferred to trust and be taught of brother man, "while the beast was taught of God." Till therefore he dethrones the authority of man, and seeks wisdom by prying for the sources of natural law in nature herself, and applying it properly, his progress will only be obtained, in the future, as it has been in the past by an endless series of blunders, impediments, and failures.

Let us now give an opinion, of the manner in which the materials comprised in a man's body, form mind, and how this mind exhibits itself.

A literary man when starved, or hungry, does not feel in a humour to write, and if obliged to do so, it is accomplished in anything but a satisfactory manner. A man who is sick or feverish cannot do any mental work. When we eat too much we feel drowsy, and disinclined for literary labour, but allow two or three hours to elapse, and we are ready for such work. After a smart walk or early in the morning, we are inclined, and ready, for anything that demands attention.

If we drink alcoholic spirits, some evil influence usually ascends to the head, and disarranges our faculties; while again, some men can only write or speak with effect, when half intoxicated. Many men also when sober, will forget everything they did when intoxicated, but remember the circumstances when they become inebriated again. We have read of the mind acting in a similar manner with people under the influence of chloroform. We would gather from these facts, that the mind is influenced in a great measure by what the body eats and drinks. Some might thus say that the intellect is in the region of the stomach, but such is not strictly correct.

We may lay it down as certain, however, that the animal mind of man—like that of the beast—is formed from the properties of the matter he eats and drinks; and man's divine mind is a property acquired by matter, after it has been transformed within him.

But the properties of matter in food, as they lie on the

dinner table, are not in a fit condition for our purpose ; neither are they while dissolving in the stomach.

Where then is the mind ?

During the dissolving process, a gaseous force is generated from our food—as more fully described in another chapter—which ascends to a space at the top of the head,—the purpose of which space, physiologists have long and vainly endeavored to discover—and there, the mind, we believe, assimilates and arranges the materials to be stored away in the memorial chamber.

Where then is the memory located ?

We see numbers of landscapes, paintings, objects, faces of men and women, etc.—and can recognise a large number of them when we see them again, or recall them in vision. Where are these memories kept, to be exercised at the will of the mind? We hear, read, and are taught, numberless words and facts from books, and can bring them up again at a moment's notice for use. Where are all these facts garnered and stored away ? We have great powers of compression, and can contract an immense landscape into a picture the size of the pupil of our eye, but to imagine that the contents of hundreds of books, with other knowledge, is stowed away in the interior of the head, seems an impossibility ; yet we can scarcely think otherwise. For although we read and hear a vast amount of information every day, most of it is soon forgotten. Those books also which are last read, are best remembered, and only those pictures we last saw, or the landscapes we viewed lately, can we recall to mind distinctly.

The brain is therefore, as it were, like a gallery of transparent pictures, each distinct class of knowledge having a section of its own ; the largest being that devoted to the subject which we study most. The last picture is always hung on the top of the preceding ones, so that unless we take care to make our pictures clear, and distinct, when photographed on the brain, and often renew such as we wish to remember, we are apt to lose them altogether.

CHAPTER IX.

CHEMICAL ACTION.

STEAM BOILER EXPLOSIONS.

A knowledge of chemical action required.—Nothing known about it by scientific writers.—Prof. Grove.—Chemical action only one form of atomagnetism.—Great separater.—Attraction the great builder.—Repulsion the great designer.—Chemical action the great destroyer.—How sugar dissolves in water.—How a nail dissolves.—Concentrated acid not so good a dissolver as diluted acid.—Soda powder. Sulphuric acid.—Amusement for speculative philosophers.—How water evaporates.—No latent moisture in the atmosphere.—No latent dryness in the sea.—STEAM BOILER EXPLOSIONS.—Facts connected with explosions.—The materials dealt with.—The manufacture of hydrogen gas.—What the United States Commissioners on explosions have discovered.—How explosion takes place.—Not by pressure.—Mingling of gases.—Prevention.

Before commencing to state what heat, light, sound and colour, are, we propose to give a definite idea of chemical action, for in this lies partly the root of our system; and, if we can convey our meaning in such a manner as to be understood, we have no fear for the other subjects to be considered.

It may be stated that, hitherto, no man has properly understood what chemical action is, although he sees evidences of its working every day, in food digesting in the stomach, in a nail rusting in water, in a fire burning, and in numberless other ways.

Prof. Grove says:—"We have no knowledge as to the exact nature of any mode of chemical action, and for the present, must leave it as an obscure action of force, of which future researches may simplify our apprehensions."

Bearing in mind the law of atomagnetism—the attraction of like atoms, and the repulsion of like poles—all mystery is at once removed from chemical action, for it is only one of the modes in which that law works.

In all chemical action the two classes of matter are present, and in all its operations, there is a constant endeavour on the part of like atoms to associate, or coalesce, with each other, and thus to be ever wearing away all material bodies, wherever situated. Thus the granite rocks on the moors have their sharp corners rounded, and their rough surfaces smoothed, by its operation. Thus a tree—when its life giving impulse or life power has left it—is attacked by the destroying force of atomagnetism or chemical action, and it is soon reduced to the dust from which it sprung. Thus also with a dead animal, chemical action at once attacks and soon decomposes it; but life giving atomagnetic action also returns, and innumerable living organisms are the result; to be again succeeded, when the supply of food is exhausted, by a different chemical action, when all the animal substances are finally dissipated, into their original mineral and vegetable elements.

All dissolving processes, such as in a galvanic battery; all cooking, or digestion of food; all transformations from one substance into another; are examples of chemical action.

All life and growth might be said to be examples of chemical action also—for this is only a sample of the same law working under different conditions—but we wish it to be understood only as the great dissolver, because the varied actions of the one great law, will be more readily comprehended, by separating the different modes of force as much as possible.

Thus if a division could be said to exist at all, we would consider the attractive force as the great builder—the repulsive force as the great designer; while chemical action would receive the ominous appellation of the great destroyer.

By way of illustration, we may offer a few examples of the particular action referred to.

Suppose we drop a piece of sugar in a glass of water, in a short time it dissolves. There is in the process no perceptible disturbance, but, on tasting the water, it will be evident a change has come over it. How did this take place? Merely by the mineral atoms in the water, finding some other mineral atoms in the sugar, and attracting them; and by the vegetable atoms finding similar vegetable atoms and attracting them. But in attracting each other, similar poles often come in contact, so that great repulsion must continually ensue. It is evident, therefore, that in every solution, there is a continuous motion and reciprocation between the atoms. The more evenly divided the two classes of atoms are in the solution also, the more active will the interaction be. Thus if we place a nail in water, it takes a long time for it to rust or dissolve, for the hydrogen, or mineral atoms, in the water are few, compared to the oxygen or vegetable atoms. But pour some acid into the water, and the chemical action is instantly increased, and the nail dissolved. Suppose we place a nail in concentrated acid, the action is, strange to say, also slow; but pour in some water to weaken it, as some would think, and the action is again increased.

Chemical action is also exhibited in various forms. For

instance, when a soda powder is mixed, great effervescence is the result shewn; while in mixing sulphuric acid in water, great heat is the result for a few minutes, without any visible motion. Why the one should have great motion, without heat, and the other great heat without apparent motion, are problems that we can only solve by saying, that all individual atoms and classes of atoms, have properties of their own, which they have possessed from the beginning, which are inseparable from them, and which cannot be annihilated. Why certain atoms are only effervescing, and others only heating, we would suggest as a capital and safe amusement, for those speculative philosophers, who are continually indulging in those harmless and incontrovertible calculations anent the end of the world; as to whether it will be caused by a comet, or a tidal wave, or a collision with Hercules, or by the extinction of the sun, or by the destruction of vegetation by insects, or by the spontaneous combustion and explosion of the earth from its internal fires. For on the acceptation of our system of matter and force, and its evident connection with all the phenomena of the heavens and the earth, their occupation in that direction will be entirely gone.

Let us now take a simple experiment of chemical action by the atmosphere. Allow a glass of water to stand on a table a whole day, and when any one attempts to drink it, he will immediately say it is not fresh. A change, therefore, must have taken place in the water. What was the nature of it? Simply that the atmosphere being composed of the same materials as the water (only in different proportions) the atoms in one, attracted the atoms in the other, and reciprocated, so that the water lost some of its own particular atoms, and became possessed of the composite particles of the atmosphere. Thus it is that water kept in a room tends to keep the air pure; not mechanically, but by its natural or chemical action with the atmosphere.

Again, if we lay aside a loaf of bread it will be chemically changed in a few days, and become hard and stale, instead of soft and moist. Scientific works tell us that there is moisture in the air, but all experience proves to us rather that there is none, for we can keep nothing with moisture in it for any length of time. Our bread becomes stale, our wood loses its sap, our gardens thirst for rain, and even our tumbler of water eventually evaporates into air, in the atmosphere.

There is no more latent moisture in the air, than there is latent dryness in the sea; and we would just as soon expect a piece of new bread to remain moist, on a warm summer's day, as a piece of stale bread to remain dry when immersed in water.

But there is no need for enumerating further examples, for every natural change or action is chemical action, and is caused by the reciprocal attraction and repulsion of the two classes of atoms.

It may not be out of place to record here, an example of chemical action, which is causing the destruction of life and property almost every day. We refer to

STEAM BOILER EXPLOSIONS.

No clue seems as yet to have been found elucidating the cause of explosion. Boilers which have been tested by an hydraulic pressure of 100 lbs. to the square inch have exploded at a pressure of 30 lbs. and although it has been surmised that this could not have been caused by pressure, and that some other force must have been in operation; yet with all the skill and knowledge of the scientific world it has evaded their research.

In the United States, Congress appointed commissioners to investigate the cause of boiler explosions, and allowed them $50,000 for making experiments, but after expending half the amount, the only discovery reported to be made, is, that the steam guages in use are very incorrect.

Before showing how chemical action is their cause, we will present a few of the facts connected with boiler explosions, culled from numerous accounts of them in the newspapers.

1. New boilers are more frequently found to explode than old ones.

2. The explosion of a steamboat or locomotive boiler almost always takes place on its starting, after having been waiting for some time. Our readers may remember the terrible Westville explosion of one of Com. Vanderbilt's Staten Islands' steamers, which was just leaving the wharf at New York, when the boiler burst.

3. The explosion in a stationary boiler takes place, when what is called a heavy pressure of steam is put on.

Let us now examine the nature of the material we have to deal with. In the first place there is the boiler made of iron. An old boiler has generally an inside coating of rust or sediment, which is more or less composed of vegetable matter, caused by chemical action between the water and the iron; whereas a new boiler has a surface of pure metal, together with the scrap iron from its manufacture.

The water which is used in the boiler is composed of two gases—oxygen and hydrogen—in certain proportions. It is well known also that if those gases are mixed in a certain but different proportion, an explosion will take place. Steam is the form of gas which rises from water when at a temperature of 212° F. As water is composed of two gases, it of course follows, that steam must be composed of the same. If we take this steam and allow it to pass through a red hot iron tube, the gases separate; the oxygen being absorbed by the hot iron, while the hydrogen, if allowed to escape through a jet, and lighted, will burn like ordinary gas. It is thus that hydrogen gas is produced.

Another fact about steam is, that it cannot be much compressed, as may be inferred from observing the lid of a kettle rising by its force; and hence comes the absurdity of trying to obtain and utilise an excessive pressure of steam.

We think we have now facts enough before us, to consider and explain the cause of explosions. Let us suppose, for the sake of clearness, that the iron boiler is transparent, as a glass bottle, or retort. Let it be half full of water. Put fire under it, and after a time the water begins to boil at the bottom, and change into its gases, which rise through the water and fill the upper chamber. There they remain transparent, till they issue from the boiler into the atmosphere, with which they chemically unite in the form of cloud.

Suppose we now shut off steam. The gases cannot be compressed, and the water cannot be made to reach a higher temperature than 212° F. without changing into steam, although the latter may be raised to a temperature of 600°. As the fire is still burning at the bottom of the boiler, the water then must be continuously changing into steam, and as it cannot gain the surface, because the upper part is already completely occupied by the gases, it must remain at the bottom. The water then is gradually being suspended above the bottom of the boiler, diminishing in quantity as it boils, while the bottom of the boiler and the steam are both increasing in temperature.

As there is thus a space filled only with gases between the bottom of the boiler and the water, the boiler must at that part soon become red hot. We have already found that steam, in passing through a red hot pipe, has the oxygen absorbed from it. The same result follows here; the red hot metal "absorbs" chemically all the oxygen gas derived from the hot steam, and at the same time throws off its mineral or metallic (hydrogen) gas, which takes the place of the oxygen until the whole becomes in consequence only hydrogen.

We have thus in the boiler, the hydrogen at the bottom, the water in the middle—suspended—and the steam at the top. But as in water there are eight parts of oxygen to one of hydrogen, it follows that the steam at the top, being the same in proportion as the water, is almost wholly composed of oxygen gas. We have stated before that oxygen and hydrogen, when separate or alone, are incombustible, but when mixed in certain proportions they form, or become, one of the most explosive compounds. It follows, then, that should the hydrogen at the bottom, and the oxygen at the top, come into contact, the explosion that would ensue will suffice to rend the strongest boiler ever made, and the greater the resistance offered, the greater the explosion and destruction would be. The materials being in the position indicated, suppose we open the valve, or start the engine, the pressure is relieved on the top by allowing the steam to escape, but what is the next result? The water immediately, by its "gravity," falls to the bottom, the hydrogen gas is forced up, the gases intermingle, and, as the opening is not large enough for the steam to escape in time, an explosion takes place by the consequent natural combustion of the gases, and then the terrible destruction and loss of life we so frequently read of, ensues. This then illustrates in a simple way what happens in the case of all boiler explosions, and their prevention can be at once made certain and simple. As steam cannot be compressed—its force being caused by its escape or condensation—an extra quantity or superabundance, is no more necessary than food for an over-loaded stomach. Therefore, the engineer should *always allow the steam to escape, after the water and steam are brought to their highest temperature.*

If by any means the steam should have been accidentally shut off, then the only safe way to prevent explosion is to allow the fires to go down, so as to permit the steam to condense into water before being agitated; or else to have an escape at both the top and bottom of the boiler.

The reason old boilers are not so apt to explode, is that the "rust" at the bottom of the boiler, being largely composed of vegetable matter, preserves the iron from any rapid chemical action, and so prevents an accumulation of hydrogen gas sufficient to cause an explosion. We frequently hear, however, of those old boilers *bursting* in a quiet, peaceable way, and washing out the fires, but doing no serious damage. Some of the plates have then been usually found to be so completely oxydized, that they were not of sufficient solidity to stand repairing.

Should the facts here stated be widely known and acted upon, we might guarantee that accidents from boiler *explosions* will almost entirely cease.

CHAPTER X.

HEAT.

Heat the result of chemical action between certain classes of atoms.—Dynamical theory of heat. Motion.—Tyndall on heat.—Christopher Columbus and his followers.—No ambition among scientific men.—Heat produced in three ways.—Natural heat.—Combustion.—Friction.—Ice melting.—Hot springs and Geysers of California.—Volcanoes caused by chemical action.—Why coal burns.—Poker experiment.—Conductive power of heat.—Tyndall's experiments.—Laboratory experiments incorrect.—Atomic action likened to a gossamer thread.—How the Crusade against the present system of science will be conducted.—Ruskin's crusade against Renaissance Painting, and Architecture.—Grove and Lardner.

FROM what we have stated in the previous chapter, it must be evident, that heat is the result of chemical action between certain atoms of the two classes.

Because motion in most cases will produce heat, the originators and followers of the Dynamical Theory of heat, assert that motion is the primal cause of heat among all atoms.

Professor Tyndall in his lecture on "Heat as a mode of motion," says:—"Heat is not the clash of winds, nor the quiver of a flame, nor the ebullition of water; all these are mechanical motions into which the motion of heat may be converted; but heat itself is a molecular motion, an oscillation of ultimate particles." But the particles are not always oscillating, there must be a force which initiates the motion.

We admit that the discovery which elicited the connection between motion and heat was a great one, but it much resembles that made by Christopher Columbus when he landed on one of the West India Islands, and went home proclaiming he had found a great country; while the vast continent of America lay still beyond his ken.

If our scientific men possessed half the daring displayed by the followers of Columbus, they would have searched into this theory of motion, indicated by Mayer and others, and by discovering what caused it, a great deal of useless writing, would have been spared the world. For there is this bane connected with science, that while we have trumpeters and drummers innumerable, we have no real leaders in it at the present time. There is not sufficient ambition among scientific men generally, to make researches, and hew out paths of their own, they would rather, like most of our professors and teachers of science, be flunkies to an eminent man's opinion, than ride in carriages of their own.

Heat is produced in apparently three different ways, but we show that the action of matter is the same in all.

Firstly, Natural heat, caused by the natural motion and reciprocal action of atoms, as explained in the preceding chapter on chemical action, or such as is known as animal heat.

Secondly. Heat given out by combustion, which is the result of an excess of the same action, when the atoms of the two classes are of a more favourable character to produce heat, and in a more favourable condition and position to reciprocate or act under a changed or different form.

Thirdly, Heat produced by friction, where that atomic action is induced by mechanical influences, as by the rubbing of two solid bodies against each other.

As regards the natural heat exhibited in our bodies, what causes it? The human body, like every other production of nature endowed with life, is a magnet; and as in other magnets, the force is exerted from the centre to either end. The stomach being the centre of the body—and the place where the food is transformed into blood, for repairing the waste of the body and increasing its growth—the material is forced from thence in both directions. While therefore a certain amount of animal heat is caused by the dissolving of the food in the stomach, and the action and inter-action of the atoms, it is also increased by the magnetic power of the body in forcing these into circulation. We also increase the natural heat of our bodies by taking our food warm. In cold climates, it may be said, the inhabitants could not preserve life otherwise.

Similar heat is produced by mixing sulphuric acid and water, as already explained, and by the pouring of water on lime. Great heat is the result in both of these cases without combustion. Although heat is induced by water, it is kept within certain limits of temperature by water also. Masons engaged in labour at new buildings, may be seen warming their dinners by placing their pitchers among lime and wet sand.

Hot springs and geysers, furnish other examples of the same action. At the Great Geysers in California, we find the ground all around them completely burnt, and exhibiting the most beautiful varieties of colour. Dozens of different kinds of chemicals are found all over the place. Water and steam ooze out under the feet, and from the sides of the ravine on either hand. Water also bubbles and boils in natural pots, and steam rushes out of natural funnels, with the force and noise of a steamboat whistle. This all shows the result of chemical action, by water, under the surface; for we find that the action has ceased in many spots, and every year it is travelling further up the valley, and breaking out in new places. Mount St. Helena which crowns the beautiful Napa Valley, and towers over all the hills around, was once a volcano, but its action has now ceased. Should not this lead us to believe that volcanoes also, are merely the result of chemical action; else why would they become extinct?

Let us now consider the second cause of heat in the process of combustion, or heat from combustion.

In the chapter on Matter, we stated that the mineral atoms were combustible; but minerals will not burn of themselves, for in all continued action there must be a union of the two classes. For example, sulphur will not burn while alone, but put some portion on a piece of wood, the two substances will burn or reciprocate together, and the consequence is we see the sulphur on fire. Iron will not burn if alone, but when it is in fine particles as filings, and separated, as in sprinkling them in the atmosphere over a gas jet, they are combustible, and when ignited burn more readily than gunpowder. Wood burns because there are mineral atoms in the composition and formation of the woody fibre. Wood is produced by a combination of vegetable and mineral atoms, through the agency of water which comprises the elements of both.

This shows conclusively, that there is reciprocity and motion in the process of chemical action, and of course, according to the properties of the atoms—the greater the motion, the greater will be the heat, provided the material elements and conditions are adapted to produce heat.

Mineral atoms are not all combustible in every situation and condition, for some find certain vegetable atoms with which they will specially reciprocate, and, often times, will reciprocate with no other sufficiently to produce heat.

The differing effects of the various combinations of the same material, are to be seen from the variety of displays in pyrotechnic exhibitions, induced by merely changing the conditions and positions of the material.

This is a branch of chemistry which is comparatively unexplored. The foundation, or *cause*, of chemical action being now furnished, and its mystery removed, there is abundant room here (that is in classing matter according to its nature and properties) for students, to make themselves useful, not to say distinguished in their day and generation.

A coal fire burns. Why?

Because we have in the coal the suitable material for producing combustion; that is, a certain species of both classes of atoms required. Coal is formed from vegetable matter, saturated by mineral solutions and gases. These have been produced from the dissolving of minerals in the interior of the earth, through the agency of water. These solutions and gases by reciprocating, permeate the vegetable matter until it is crystallized or petrified into coal. By the mechanical appliance of fire, or combustion, to this coal, a reciprocal action again takes place in a new and different form, between the atoms of the two classes in the coal, in consequence of which these gases and particles are again set free, or are repelled from the pieces of coal;

and by the intense action of the atoms both heat and light are produced.

If we put a poker in the fire it becomes red with heat, and, should the fire be made hot enough, it will melt. Why? Because the polarity of the atoms of the iron are reversed, and the mineral atoms of the fire have a greater attraction for the atoms in the poker than the iron has. Gradually, therefore, the polarity of the atoms is reversed, and as the fire becomes hotter, they are repelled from the poker and amalgamate with the atoms in the fire; for the greater always influences the less. The end of the poker, if held in the hand, is found to be hot also, although it is some distance from the fire. Why is this? Because the metal, being conditionally a combustible material, possesses a conductive power of heat, just as it has a conductive power of sound.

If we hold the hand before a fire we feel it warm. Why? Because certain atoms in the fire have an affinity to similar atoms in the hand, and are seeking so draw them out. If we can bear it long enough for them to do so (and our hand is burned) then we feel it just as much as if a hook were put into the flesh, and a piece were torn off.

We now come to the third division, and consider how heat is produced by friction.

Supposing we take two pieces of metal. in which, by polarity, the atoms are all lying in one way. That is the "north" pole of one atom, is next in position to the "south" or opposite pole of another, and so on all through, leaving the middle of the atom exposed on the outside of each piece. If we then rub the pieces together, we are disturbing the position of the poles by rubbing similar points together, and as similar poles naturally repel one another, an action commences in the atoms, if the friction continues long enough, to reverse the poles. Should the friction be intense, we have the metal dissolved, that is, repelled from each bar or piece, and amalgamated with that of the other.

In rubbing two pieces of wood we have a similar resultant action, but as the heat increases the atoms are thrown off in the form of gas, and such is the nature of the action in this class of atoms, that combustion is the result of their reciprocation.

Sir Humphrey Davy's experiment of dissolving two pieces of ice by rubbing them together, caused quite a discussion at one time, and strange conclusions are drawn from it by Prof. Tyndall and others. But it is easily explained. In ice there are a considerable number of mineral atoms, and the friction of these at opposite poles, produces a heat which gives the atoms of one piece, a greater attraction for the atoms in the other, than they have to the piece they are connected with, and thus they are repelled; a sufficiency of oxygen being thus set free by the heat, or being present in the atmosphere, it combines with the repelled mineral atoms and forms water.

We have now given, it is hoped, such an explanation of heat as will be understood, and, by way of contrast, we make a few quotations from our most distinguished writers on the subject.

In reading Professor Tyndall's lectures on "Heat as a mode of motion" we admired them very much, and his experiments were no doubt beautiful to behold. But, while we do not deny any fact he illustrated, or any experiment he performed, we certainly do deny many of the conclusions he arrived at. For, regarding the assistance to be derived from experiments performed in the laboratory, in corroborating assertion, or proving theory, we are very much in doubt; inasmuch as the mode of action in atoms, in all the phenomena which they exhibit, are biased by the slightest influences, to an extent that few of our physicists even dream of.

In all natural action, the atoms, in order to assume their atomic position—that, is, regarding their polarity to each other, which affects their form and disposition, their transparency and opacity—must have perfect rest and an even temperature. The typical form of atomic action, may be likened to a glistening gossamer thread springing from earth to heaven, and which if undisturbed, becomes as an endless gleam of silver, straight, pure, and transparent; but it is so sensitive to motion, that the slightest zephyr that blows, will twist and turn it, and cause it to become knotted and ragged. Laboratory experiments are therefore like unto the last, for the action being in nearly all instances forced, the chemists see only these knots and rags, and argue from them. We believe, therefore, that more good would result, and more truths would be discovered, if physicists would appeal more to nature, and confine themselves mainly to a scrutiny of experiments conducted by nature herself, in her Alpine Glaciers, her stalactitic grottoes, and her forest dells of ferns and flowers; for every one could then test for himself the truth of the theory propounded, without the aid of expensive and delicate apparatus.

While, however, we object to the common practice of theorizing, and basing grave laws of science on forced results obtained from in-door experiments, we have no objection to their endeavouring to prove by such tests, the truth of problems found and solved in nature, yet we fear that our physicists generally proceed in a contrary manner. The crusade against the present system of science therefore, will have to be conducted on much the same principles, and for somewhat similar reasons, that Ruskin waged war against the great Renaissance revival in painting and architecture; by appealing to nature as a paramount authority in everything, over the opinions or doctrines of any man, or class of men, however eminent or distinguished.

A number of persons again may read the same sentence of a foreign language, yet scarcely any two will translate it similarly: and, as a person who is ignorant of the language could only guess the meaning of the sentence, so one who is ignorant of the composite law of nature and its working, can only stumble at the processes of any of its phenomena. It may seem a startling thing to say, but we assert that neither Professor Tyndall, nor any other man, can read nature, or experiments in nature or in the laboratory aright, until they understand magnetism and atomic action.

For instance, he says:—"A leaden bullet in hitting a target is much hotter than an iron one, for iron has a greater capacity for absorbing heat than lead." This is manifestly incorrect, for such heat is entirely induced or caused by friction. The ball of lead in hitting the target is crushed out of shape; the intense friction of the particles consequent on this circumstance, must produce great heat. The harder iron, on the other hand, by not yielding, has little or no friction, and of course very little heat.

Again, from this illustration, they argue that if a solid body be stopped, a certain amount of heat would be generated, according to the rate at which it was going. But it is obvious that if the same leaden bullet had been stopped in sand, so as to preserve its shape, it would have generated no more heat than the iron one.

Thus they argue, "if the world was suddenly stopped, enough heat would be generated to reduce it in great part to vapour." If the earth were stopped by coming into collision with another world (an absurdity also) and it was crushed out of shape, we could account for the friction causing heat, but the simply stopping its motion through space would not do so.

In order to carry out this great discovery to its fullest limits, according to Dynamical principles, we are also told by Prof. Tyndall that "the earth on stopping would assuredly fall into the sun, and the heat generated by the blow would be equal to five thousand worlds on fire!" In order to calm the fears of unsophisticated mankind, we will show in another chapter, how, under the existing laws of nature, such an event is impossible.

Prof. Grove, while endeavouring to show that heat is motion, feels himself compelled to say: "We know not the original source of terrestrial heat, far less that of solar heat."

Brewer states that the principal source of heat is the sun, and Tyndall also means the same thing when he says: "We are all souls of fire, and children of the sun."

The climax is, however, reached by Dr. Lardner when he says: "Heat is propagated by radiation, which is apparently independent of matter." If the Professor could point out a place where there is heat, or anything else without matter, the scientific world would be largely indebted to him.

CHAPTER XI.

LIGHT.

Light caused similarly to heat.—Propagated differently.—Three divisions. —Light without heat.—Light with heat.—Propagated light.— Auroras explained.—Phosphorescence.—Tyndall refuted on molecular motion.—Fire-flies.—Lighting gas by the finger.—Auroras from trees.—Candle a guide to light.—Four things required to be looked at.—The flame.—The heat.—The light.—And light as an object.— All light, reflection.—The undulating theory disputed.—Light instantaneous.——Light cannot travel half a mile.—Sight travels 286,000 miles a second.—Flame not seen in daylight.—Astronomical fallacy of star light.—Undulation follows Emission into oblivion.— Tyndall's security for the continued acceptance of the Undulatory theory, overthrown.

In considering this subject, we find that light, or flame, is produced in a similar manner to heat, but it is transmitted differently. If we are right in stating heat to be the result of chemical action, between two certain classes of mineral and vegetable atoms, then light is produced in the same way; and, in most instances, we cannot get one without the other. Light however does not always indicate, or give heat, and some materials do not afford such a bright light as others. This is owing, we believe, entirely to the proportion, condition, and position of the materials yielding the light. For example, a lamp may not give a very bright light by itself, but put a glass shade over it, to secure an oxygenized current around it, and we have a light equal to gas.

We consider light under three divisions.

First, light without heat or combustion.

Second, light with heat and combustion.

Third, how light is propagated.

In the first division we have the aurora, fire-flies, phosphorescence on water, etc.

How are those lights caused?

Merely by two classes of suitable atoms, combining in suitable proportions, and reciprocating, causing friction and light. The aurora is caused by the mineral gases from the poles, or in polar latitudes, mingling with the vegetable gases from equatorial, or warmer latitudes, in certain proportions, and under certain conditions. These same gases in other proportions and conditions will produce rain and snow, lightning and thunder, hail and storms; so that it will be observed, nearly everything depends upon proportion, condition and position.

On the Pacific ocean, we have seen the waves, made by the steamer, rolling in a volume of liquid fire, while a long stream of light was left behind almost bright enough to read by; yet there was no more heat in it, than in the surrounding water.

Why was this? Because the atoms were not in a position to produce and show a development of heat.

If these light giving principles could be extracted from the water, we have no doubt they might be placed in a position to exhibit heat also.

If merely cutting through the water lights it up, it may be asked why does the water not always shine, seeing that the light producing properties are always present in it? Because it is not in a condition to do so. *Light is caused only when the atoms are in excessive motion, when the poles are disturbed and clash with each other; and they are only in this condition when the atoms are unequally divided.*

In calm water, the different classes are thoroughly at rest, and arranged in natural atomic position, but the crest of a wave in falling, or a steamer in motion, presses the atoms into collision with each other, many consequently combine, while instant commotion ensues throughout the mass, the greater lights attracting the lesser; till by rapid attraction and repulsion, the atoms are in order again, and darkness reigns supreme.

It may be asked why these little globes of light, having once been set in motion, do not continue to attract the light producing atoms, till they form an immense ball of fire? Because crystallization, or the aggregation of any particular class of atoms, can only ensue when such atoms preponderate over others, and as the light atoms in the water, are few, in comparison to the mass of other atoms in solution, therefore, as before, the lesser must necessarily give way to the greater.

The streak of light left by a fish as it dashes through the water, or a meteor in passing through the atmosphere, is evolved on the same principle. (See chapter on Auroras.)

In the fire-fly, the light is caused by the atomic mingling of two different kinds of suitable material, one of which the fly manufactures and developes in its body from its food; the other is probably derived from the atmosphere. The heat from this light is very slight, no greater than the temperature of its insect body.

Many healthy persons after a brisk walk will ignite our ordinary coal gas, by merely applying their finger to the jet.

This is caused by the animal, or concentrated vegetable gas, from the body, reciprocating with the mineral gas from the pipe.

A light is often seen at the ends of branches of trees on a cool morning, resembling miniature auroras in appearance, this is caused by the vegetable gas from the tree, reciprocating with the mineral atoms in the air.

The second and third divisions on the subject, viz: light with heat and combustion, and how light is propagated, we will consider together. The explanation of daylight, we will reserve for another chapter.

Taking a candle for our means of illustration, we find that there are four things required to be looked at.

First, the flame itself.

Second, the heat from it, extending twelve inches or more all around.

Third, the light from it extending in a greater or less degree for ten or twelve feet.

Fourth, The phenomenon of the light itself, looked at from a local distance of a few miles, according to the state of the atmosphere.

In the first, we burn our fingers if we put them in the flame.

In the second, we feel warmth but we are not burnt.

In the third, we do not feel the heat but the light is bright enough to read by.

In the fourth, we lose both heat and light, locally, and can only see the flame as a distant object.

The phenomena felt and observed in these four instances, must of necessity make the action in each several case different from the others.

Now what is the action?

In the first, we have the atoms of the materials under intense motion and friction, causing heat and light.

In the second, we have the same action, but with differing materials, or atoms in a different condition and in a less degree.

In the third, as the light is too far removed to be apparently affected by, or to take part in, the chemical action going on in the flame, the cause and effect must be ascribed to something different. This we assert to be due to the reflective power of the atoms. If materials can combine into a flame, and reflect a powerful light all around from the action of the atoms on one another (for all light may be said to be reflection) and if all substances reflect light more or less, the atmosphere being composed of the same kinds of material, must be possessed of the like reflective power. Consequently this reflection of one atom on another, produces what we experience as light, the intensity of the light or reflection decreasing, as the distance from the flame is increased.

In the fourth instance, we have a pure case of sight; that is, we are looking through a transparent medium at a mass of reflecting atoms; but the scientific world have invented a theory connected with it, which is worth combating.

Prof. Tyndall, in his lectures on heat, says:—"Sir Isaac Newton supposed light to consist of minute particles, darted out from luminous bodies;" this was the celebrated *Emission* theory. To Dr. Thomas Young, however, belongs the immortal honour of establishing on a safe basis the theory of *Undulation*. According to the notion now universally received, light consists first of a vibratory motion of the particles of the luminous body, which motion is communicated to the ether in which they swing, and thus it is transmitted in waves to the eye.

The act is as truly mechanical as the breaking of the waves on the shore." The idea is a very beautiful one, but very hard to comprehend, for he adds, "a wave of light comes from Jupiter to us in a second! 186,000 miles." Again different waves of light have waves of different lengths. " 39,000 waves of red light placed *end to end* make up one inch; accordingly 470,439,680,000,000 red waves enter the eye in a single second; while 699,000,000,000,000 of violet waves enter in the same time."

A great mistake, it will be observed, is made between light and sight.

Both Light and Sight are instantaneous. If we shut our eyes and open them again, we can see any object within their range, whether five or ten miles away. It does not require a distant vessel or house to send waves of their colour to us in order that we may distinguish them; so also is it with a candle.

If we place it in a room, we see it as a bright light, and everything in its immediate vicinity is rendered distinct by its reflection; but place it a hundred yards away, and we only see it as a point of light. Place it ten miles away and we cannot see it at all, and if it be kept burning for a million of years, the light will never reach us.

It is asserted as one of the poetical facts of astronomy, that the light of probably some thousands of stars has been travelling towards us ever since, if not before, Adam was created, and it has not reached us yet. If, as we assume, sight and light are one, both instantaneous, then we may safely affirm, that the light from these thousands of stars beyond our vision, will never reach us.

Again, if we hold a piece of glass before us we can see Jupiter through it. The waves being communicated through the ether, which is *everywhere and in everything*, glass included, of course it goes through it to our eye. But suppose we hold a piece of cloth before us—which is much more porous than glass—we cannot see Jupiter. *Waves* and *ether*, therefore, have nothing whatever to do with seeing the star, but sight and a transparent medium, everything.

Lastly, if a fire is lighted a mile or two away during the day, we see only the smoke; the flame is not visible at all; while at night the light only is visible. What is to be said of the waves of light during the day?

From what we have remarked, light at a distance must be regarded only as a question of sight, and the Undulation theory will have to follow its predecessor Emission, into oblivion.

Prof. Tyndall, in his lectures in New York, stated, that while it was quite logical for any one to insist, that sufficient evidence might be brought forward in the future, to overthrow the undulatory theory of light, just as other theories as widely accepted had been given up before, yet it was as unlikely to be overthrown, as the theory of gravitation. If our theory is correct, —which we are endeavoring to prove—then the foundation of security on which he has built his hopes, is gone, for we adduce enough proof in another chapter to insure its rejection.

CHAPTER XII.

THE SUN AND SUNLIGHT.

Professors Thomson, and Tait, on the Sun.—The Sun a huge furnace.—Herschell on the waste heat of the Sun.—Temperature of space.—Our view of the universe.—The solar system an inhabitant of it.—The Sun a stomach.—The atmospheres, the flesh and bones of the solar system—Movements regulated by Magnetism.—The Sun an inhabited world. How Sunlight is caused by Magnetism.—The Sun, Earth, and planets, Magnetic batteries.—Sun the main battery and head office.—Planets telegraph stations.—Sunlight caused in a similar way to the spark at the poles of a battery.—The "Journey to the Sun."

Having explained how ordinary lights are caused, and what our views of light are, we now proceed to discuss the character and phenomena of daylight.

We do not profess to be astronomers, and would not wilfully run counter to the grand discoveries which eminent men such as Newton, Kepler, and Herschell, have made in connection with astronomy, but as our system clashes with theirs, and as it is impossible for us to see how daylight is caused by a huge roaring furnace; it is essential that we lay down a system, in accordance with natural law, as understood by us.

Before doing so, let us examine a few of the statements made by teachers in physics. Professors Thomson and Tait, in an article on "Energy" in "Good Words," 1862, give four theories on the nature and action of the sun. The fourth, which they say is probably the true explanation, is as follows:—" According to this theory, matter when created was diffused irregularly through space, but was endowed with the attractive force of gravitation, by virtue of which it gradually became agglomerated into masses of various sizes. The temperature produced by collision, etc., would not only be in general higher for the larger bodies, but they would of course take longer to cool; and hence our earth—though probably in bygone ages a little sun—retains but a slight amount of its original heat, at least in its superficial strata, while the sun still shines with brilliance, perhaps little impaired. Supplies of energy are, no doubt, yet received continually by the sun, on its casual meeting with masses traversing through space, or the falling in of others revolving about it; just as on an exceedingly small scale, the earth occasionally gets a slight increase of *kinetic* energy, by the impact of a shooting star or aerolite." " But it is not probable that the sun receives in this way more than a very small portion of his heat. He must therefore, at present, be in the condition of a heated body cooling." " But it will take seven thousand years, before his average temperature can go down one degree!"

This conclusion is a very safe statement to enunciate, but we are astonished that any man with an established reputation should make it.

The accepted idea, therefore, is that the sun is a huge furnace, which is being continually fed with fuel to keep the heat up. This heat, we are told, is given off in every direction, and as the planets—which the sun lights—are mere specks in comparison with the vast open spaces between, there seems to be, as Sir John Herschell says :—" An enormous waste or what appears to be waste."

"Take all the planets together, great and small, the light and heat they receive, is only one 227 millionth part, of the whole quantity thrown out by the sun. All the rest escapes into free space and is lost among the stars, or does there some other work that we know nothing about."

This very fact should have told Herschell the theory was a bad one; for as there is nothing more perfect than nature, in all its inherent principles or laws; so it is not in accordance with any of God's works, that such a furnace should be poised in the heavens to light and heat a few planets, when it could equally perform the same service to a million of them.

But there is another strong objection. Herschell says the temperature of empty space is no less than 230° F. Thermometer below zero." If the sun is a furnace, then according to all rules of furnaces, the heat must be all wasted or absorbed by this cold region of space. Or, if it is really the heat of the furnace, which we feel on a summer's day, then the "empty" space must be hotter still, for the nearer we approach a furnace, the hotter the temperature becomes. But the reverse in this case is the reality, for the higher we rise from the earth, the colder it becomes, and the mountain tops are covered with ice and snow, all the year round.

We must, therefore, give up the supposition that the sun is a furnace, and seek for an explanation of the light, and heat, which we receive from it, in the theory of atomagnetic action.

In giving a theory of the sun, we are obliged to lay down a theory of the universe, which, it is unnecessary to say, differs from that accepted by astronomers.

Our universe is a vast body of which the solar systems are the inhabitants, and yet these are all so regulated in their movements, the one with the other, that even the universe itself, as far as we see it, may be but as a solar system, in a universe more boundless still.

Our solar system then, may be compared to an animal body with a stomach—the sun and the planets being different members of it, all having regular arteries and channels to travel along, and from which they cannot deviate. And just as the different members, or parts of our bodies, are connected with blood, flesh, and bones, so the planets are all connected with the sun, and with each other, by their atmospheres; a firm, although elastic, and invisible material.

This accounts, firstly, for their position.

Secondly, their movements are regulated by the law of MAGNETISM.

We have shown that every atom is a magnet, every conglomeration of atoms, therefore, must also be a magnet.— The sun, planets, moons, and comets, are therefore magnets; and they act upon, reciprocate with, and attract and repel each other. Thus it is that sometimes the earth approaches the sun, and again recedes, yet still keeping its fixed course and position.

But the sun could not attract any planetary body close to, or into itself, for it must be like unto the earth—with its own atmosphere; every body and planet therefore, having its own emanations, or atmosphere, enables it to keep its solid parts free, and apart from every other body.

Thirdly, the life action, producing heat, light, or electricity, and all other natural phenomena on the earth, is shown to be the result of chemical action, caused by magnetism or atomagnetic action.

We do not think therefore that the sun is a furnace, but rather that it is a vast body probably inhabited like our own earth. Being the largest also, it occupies the centre of our system, and controls the life force or magnetism of the whole.

How then is sunlight caused?

As we have already said, the sun and planets are magnets. They are also magnetic telegraph batteries with their poles, and between them there is an atmosphere acting as a connecting wire or medium,—by which they all communicate with the head office, the sun.

Whatever part of our earth's surface is in line with the sun, there is bound to be a powerful reciprocal force and action between them, and the more direct and unobstructed the line is, the greater will be the action.

How do we know this? By the phenomena of sunlight. Supposing for illustration we cut the wire of an ordinary telegraph line, and bring the two ends into merely the slightest contact, we have a spark of light or electricity—Why? Because the force of the decomposing battery, is transferred from its poles within the solution, to the partially disconnected poles on the line, inducing them to attract and unite; this force is so concentrated by the slight contact, that minute particles of mineral matter are thrown off the wire, by reciprocating with the particles of oxygen and hydrogen in the atmosphere, and thus light and fire, or electricity, is the result.

Sunlight, as we will now show, is an exhibition of this spark on a grand scale.

The earth, as a battery, is continually, by the action of water in its interior, dissolving and reforming minerals, and throwing their particles into the atmosphere. But on the surface the vegetable atoms assert themselves,—the conditions, heat, light, and moisture, being favourable for their development—and thus they grow into vast forests, and these in turn produce and feed animals. The emanations from these forests, as well as other vegetable emanations, are sufficient to form a rim or covering of oxygen all around the earth, in close contact with it.

The sun, we believe, is of an entirely similar formation, with its dissolving and reforming action, its vegetation and animals, and its oxygen atmosphere.

The mineral gases rising from both sun and earth being lighter than oxygen, occupy the higher spaces of the atmospheres. The connecting medium—called by the supporters of the Undulatory theory of light, "luminiferous ether,"—all through space, must necessarily be of the same nature, because the conditions for the presence and maintenance of vegetable atoms do not exist there. The earth and sun, being thus like working magnetic dissolving batteries, connected by a metallic medium, there is a strong action existent between them, as between two poles. But the rim of oxygen at either end, partially breaks the connection. The result then is the continuous spark of electricity, on a scale, commensurate with the great size of the batteries—for the force of the sun and earth striving to meet each other, through the hydrogen or metallic medium, causes the elementary atoms of oxygen and hydrogen to reciprocate in the lower atmosphere, and, as may be shown by experiment, the combination of the two gases with the intense friction and motion between the poles and atoms, produces our glorious sunlight.

A similar action goes on in the sun, and thus its brightness is accounted for. But, while we have light and darkness every day, the inhabitants of the sun may not know what night is; for owing to its central position, it is doubtless at all times reciprocating with planets on every side of it.

It may be said by some opponents, that the electric spark burns us if we touch it, why not daylight too? Because usually, the action is so diffused, and the continuity disturbed by winds, etc., but if we concentrate the light on the hand with a lens, it may be burnt also; besides how many people have their faces sun burnt by direct exposure to the sun's rays?

If we ascend out of this oxygen atmosphere, we gradually enter the cold region which æronauts have experienced. Why? Because we lose the region of the compound which causes the development of light and heat.

It may be asked why the moon does not reciprocate with the earth in a similar manner to the sun?

Because very probably by being connected with us, its atomagnetic character resembles our earth's, and is of a similar pole, consequently it would not act with us; but it also reciprocates with the sun.

Daylight, then, is a vast process of chemical action between the atoms of the atmosphere, induced by the reciprocal magnetic force of the bodies of the sun and earth; this is the reason exposure to sunlight in temperate latitudes is so healthy, for the intense action and motion in the atmosphere, must produce a corresponding action and circulation in the bodies that are exposed to it.

Thus we have given an explanation of the general features of that wonderful orb, which shines on us with such splendour, and which regulates our movements in a manner that the simplest may understand; and yet proved on our atomagnetic theory to be governed according to natural law, which is infallible. Thus have we done away with, it is hoped, forever, that superstition worthy only of untutored minds, which imagines the sun to be a huge furnace, feeding upon comets and shooting stars, and now and then swallowing a planet to assuage its hunger. Thus have we done away with the necessity of thinking so little of the works of nature, and so much of ourselves, that a few planets should monopolize a great furnace, which—if it had them in proper position—could light and heat up all the starry hosts of heaven. And thus, in conclusion, are turned into ridicule those theoretic descriptions of a "Journey to the Sun" by a sensational philosopher, in which the voyagers—impervious to heat or cold—after travelling far through "empty" space, at last encounter flames of burning hydrogen thousands of miles long, and see through the rifts of the raging fiery clouds, the red hot nucleus of our luminary within.

CHAPTER XIII.

COLOUR.

Undulation theory of colour—What is the force which governs colour—Primary causes overlooked as usual—Great display of Arithmetic—Looseness in Science—When we will freeze to death—Portland Scientific Convention—Tyndall on the vibratory theory—474,439,680,000,000 red waves a second—This theory questioned—No colour on the Earth—Herschell—Helmholtz—Science like a voyage of discovery—We introduce the atomagnetic theory of colour—Colour a property of matter—Colours of mineral flames—Why is the sky blue?—Tyndall's "Scientific use of the Imagination"—The setting sun red—The hills purple.

HAVING given the cause of light, and its modes of action in different forms, we now give an explanation of that element of it which we reserved for a chapter to itself.

As light has been explained by the theory of undulation, so also has colour, which in one way is inseparable from it.

The theory is said to have been suggested by the vibration of a harp string. The shorter the string is made, the greater are the number of vibrations produced by the same force; and the shriller the note becomes.

Thus it is said to be with colour. Light vibrates, but coloured lights vibrate more.

From experiments made, it has been shown that red rays are propagated the shortest distance, and violet rays the longest.

Consequently it is laid down as a settled fact, that a certain number of vibrations, or waves per second, produce red, a few more produce orange, and so on all through the colours of the spectrum.

We can understand why a harp string should vibrate rapidly, or slowly, because we know the force which is applied to it. But what is the force which controls the colours, and makes the violet to vibrate more rapidly than the red? No explanation is made of this circumstance, and as it never seems to have been thought of, the statement about the waves is a merely empyrical one. This shows again, how our leaders in science are forever striving to understand secondary causes and phenomena, forgetting altogether, or not seeming to remember, that there must be a first cause for everything. They find something, out of which by a great display of arithmetic, astounding announcements may be made, and they therefore seek to obtain all the glory possible for this discovery, before another shall find the first cause of his secondary force, and blast his fame.

From this looseness in science, we have one philosopher stating that we will all freeze to death on this earth, while the sun is gradually cooling; but it will take *seven thousand years to go down one degree!* Another states we are rushing into collision with Hercules, but we need not pack our trunks for a million of years or so yet. Thus every scientific journal we look at is filled with some extraordinary theory, and instead of being laughed at, the pedants are considered as men of genius. An exception, in this respect, must be made to the papers delivered at the late Science Convention in Portland, United States, which have been so thoroughly ridiculed, not only by outsiders, but by the Americans themselves, that we only hope it will teach them a lesson, which will not soon be forgotten, by dabblers in science all over the world.

This vibratory theory is improbable on the face of it, and seems very sensational.

Prof. Tyndall on heat says:—"Light travels through space at a velocity of 192,000 miles in a second. Reducing this to inches, we find the number to be 12,165,120,000. Now it is found that 39,000 waves of red light placed end to end would make up an inch; multiply the number of inches in 192,000 miles by 39,000 we obtain the number of waves of red light in 192,000 miles: this number is 474,439,680,000,000. *All these waves enter the eye in a single second.* To produce the impression of red in the brain, the retina must be hit at this almost incredible rate."

But all colours are not 192,000 miles away. For instance we set fire to a red light three feet from us, and in a second we perceive the colour. By Tyndall's own figures, the greatest number of waves that could possibly exist in that distance would be 1,404,000. This number then produces the impression of red on the brain, whereas he says nothing less than 474,439,680,000,000 could do it. It is evident that there must be something faulty about this theory, or it would not break down so easily.

Again, if we examine a bouquet of flowers, is our eye being hit by millions of waves of colour, coming from them? We think it more probable that our nose is being hit by millions of waves of the atoms of perfume, for we have a stronger sensation from the matter by the one, than from the *properties* of the matter, by the other.

Nearly all who have written on the subject in these latter

days, seem to assert that there is no colour whatever in the earth, and that all the brilliant hues which we see in a summer's day, are imparted to objects by light.

Herschell, Tyndall, Maxwell, Helmholtz, Brewer, Parker and others, endeavoured to prove this, and there is no doubt that it is the rock on which they have stranded.

Science in many ways is like a voyage of discovery into unknown seas and rivers, amidst which navigators have heard that certain lands are to be found. Now one philosopher and now another takes the helm, and after placing buoys at different points to mark the channels of knowledge from which no succeeding explorer must deviate, and whose correctness no one must question; they stumble along from one quicksand to another, till they are lost amidst rocks and shoals. This is what is being done in the region of colour, as well as in many other branches of science.

One observer having a slight show of plausibility to support him, asserted that there was no colour naturally on the earth, but that it was imparted to it by light. There being no better theory at the time, it was not questioned much. A buoy was immediately placed, and every scientific man guided by this, at once commenced to puzzle his brains in order to account for the colours in the light, and how they acted on objects around us. First, the refraction theory, and then the undulatory theory was started, but they have both miserably failed.

We now introduce the atomagnetic; and as it springs from *primary* causes, and elements, altogether, and as no buoys or soundings by previous observers are recognized unless they stand our own test, we feel sure it has better claims to stability than any that have been previously advanced.

As we stated in the first chapter, matter is naturally possessed of certain properties which are inseparable from it. The mineral atoms have naturally, as inherent elements, the cold colours; blue, black and white; while the vegetable atoms are naturally possessed of the warm colours, red, yellow, and orange. Of course there appear to be exceptions. Gold is yellow, but it is very scarce and prized accordingly. Sulphur is yellow also, but it sheds a blue light when burned. Sometimes we see a blue flower also, but they are very rare indeed. Every material we know of has a colouring element of its own, caused by the colours of the different classes of matter composing it. Grass and most vegetation is green—a mixture of the yellow and blue of the two classes of atoms—while all the beautiful variety of colours we see in a flower garden, are derived in a similar way.

We once saw a professor experimenting before an audience of an evening with different minerals, showing the colours which they assumed on being set on fire, and were astonished that he did not, or would not, understand what gave these flames their various colours. It could not have been from sunlight, for there was none; neither could it have been gaslight (or "bottled sunshine," as it is called by our sensational philosophers) for that was turned down. It must, therefore, as a necessary consequence, have been inherent in the materials themselves, and we cannot see how it could be properly explained otherwise.

Again, by the accepted theory, a bouquet of flowers ought to be colourless at night, but if we hold them to this mineral flame, we see the natural colours of the flowers just as in daylight. Thus proving the fallacy of the assertion that everything is colourless, and that it is the sunlight that gives them, or indeed anything else, their colour.

It may be asked, how is it we see a spectrum of colours in light at all? Because the atmosphere in which light is exhibited, is composed of all the materials—in a gaseous form—of which this earth is composed, and of course they retain their inherent colours also. Consequently light by coming through these different materials, displays also their colours to us.

If the spectroscope is held to one of those mineral flames we speak of, it shows a spectrum of all the colours too; for a similar reason, that in order to produce flame at all, there must be a mixture of the two classes of matter; and adding the materials of the atmosphere in which the experiment is shown, there need be no difficulty in collecting all the colours together. In corroboration of our system, we give a few facts illustrative of the different colours, and the materials to which they belong.

Why is the sky blue? is a question which Prof. Tyndall says is the most difficult one in meteorology; and the explanation he gives of it in his lecture on "The Scientific Use of the Imagination" is simply, to our mind, incomprehensible. Compare our account of it with his. The atmosphere nearest the earth is dense, and composed mainly of oxygen, while the higher atmosphere is rarified, and composed mainly of hydrogen, or mineral atoms. As blue is the main, or distinctive colour of the mineral atoms, the sky is blue in appearance, because in looking upwards, we are gazing through a greater volume of the mineral, than the vegetable atmosphere. But when we look at the rising and setting sun, or moon, we see them red. Why? Because we are looking through a dense volume of vegetable atmosphere, and red is the distinctive vegetable colour.

Why are distant hills purple? Because we have a blending of the two atmospheres, the blue and cold mineral atoms of the hill, seen through the red warm vegetable atoms of the valley. The chasms in the Alpine glaciers are blue, thus showing the mineral nature of ice and frost. Cobalt blue again, is highly magnetic, a characteristic of mineral matter, only in an intense degree.

Iron when it rusts, becomes red with oxide, through the action of vegetable atoms upon it.

Many more examples might be adduced in support of our assertions, but we think enough has been advanced to establish their truth.

CHAPTER XIV.

ELECTRICITY.

All light is Electricity.—Greatest Scientific delusion of the day.—Magnetism and Electricity essentially different.—Quotations to show how little is known about either.—Dr. Thomson.—Parker's School Book of Philosophy.—Sir Wm. Thomson on Electricity flowing.—Prof. Tyndall also confesses ignorance.—Prof. Grove.—Prescott's History.—Electric spark, what composed of.—No combustion without a mixture of the two classes of matter.—The cause of lightning.

FOLLOWING up our chapters on light, we come to Electricity. In fact it ought to form part of the chapter on light, for all light is Electricity. The greatest scientific delusion of the day, is the supposition that electricity has anything whatever to do with telegraphing, except, that under certain necessary conditions, the magnetic force in the wires ofttimes exhibits it.

Magnetism and Electricity are essentially two different things, but they have been so interchanged by writers, for several generations, that the public mind is uncertain where the boundary line lies.

We hope however by a few facts, and illustrations, to be able to show clearly what each phenomenon is, and the difference between them.

In the first place, to show how little is really known about either, we will make a few quotations from well known writers on the subject.

Dr. Thos. Thomson, in his treatise on "Heat and Electricity" says:—"Electricity is the property acquired by bodies, of attracting and repelling light bodies, through the action of friction on them." This is exactly the property possessed by a magnet. Why then should two names be given to the same thing?

In another chapter the same writer says:—"I shall now give an account of the recently discovered facts, which have shown the dependency of magnetism on electricity."

Then follow a number of statements which in our view would show the reverse, for the names are merely transposed. Notwithstanding all he says in his treatise regarding electricity, he has at last to confess, in speaking of the powers of attraction and repulsion:—"We are altogether ignorant of the cause of these properties."

In Parker's School Book of Philosophy we read:—"Electricity is the name given to an imponderable agent which pervades the material world, and which is visible only in its effects." Again, he has also to confess:—"The nature of electricity is unknown."

Sir Wm. Thompson in "Good Words" 1867, says:—"In every kind of electric telegraph long or short, aerial or submarine, a signal is sent from either end, by causing electricity to flow through an isolated metal wire." A very rapid and singular kind of flowing it must be; for he tells us that it flows at the speed of, "twelve times around the globe in a second." He has also to confess his want of knowledge on the subject, for he says:—"It may be regarded as probable that there is a real electric fluid, and that this fluid really flows through the wire; but in the present state of electric science, we cannot tell, or even conjecture on any ground of probability, whether the true positive electricity is that which is commonly so-called, or," etc.

Prof. Tyndall also says in his lectures on "Heat:"—"We have every reason to conclude that heat and electricity are both modes of motion; we know experimentally that from electricity we can get heat, and from heat that we can get electricity. But although we have, or think we have, tolerably clear ideas of the character of the motion of heat, our ideas are very unclear as to the precise nature of the change which this motion must undergo, in order to appear as electricity; *in fact we know as yet nothing about it.*"

Prof. Grove says:—"How the phenomena are produced to which the term attraction is applied, is still a mystery."

In Prescott's "History of the Electric Telegraph"—which is a compilation of facts from De la Rive—we read:—"The theory most generally admitted is that there are two electricities, each composed of particles that mutually repel each other." What would be the use of two electricities, having the same properties; in fact we cannot see how there could be two, if they had the same force. Moreover, two differing electricities could reciprocate, while two similar ones could not. We do not see either how a message could be sent along the wires, if the particles mutually repelled each other. If on the contrary they mutually attracted and repelled, there might be some common sense in it.

From the quotations given, it will now be seen how little is really known about either electricity or magnetism, and it is no wonder, seeing the manner in which they have been mixed up, that no one has been found capable of separating them. This however we shall now endeavour to do.

It is admitted that the *spark* at the poles of a magnetic battery is electricity, and we are assured that this spark flows through the whole length of the wire. This can easily be proved by finding out what the composition of the spark is.

There can be no light or fire without a combination of the two classes of matter. As stated in the chapter on Sunlight, the spark from the telegraph wires is composed of the two gases, oxygen and hydrogen; being minute atoms thrown off the metallic wires by the force of the magnetic action in them, and reciprocating with the oxygen and hydrogen in the atmosphere.

If then electricity is a combustion of two different materials, how could this combustion flow through a metal wire? Or how is it that the wires do not burn away instantly? Moreover the spark is only exhibited when the wires are in a certain position, and when the force is concentrated.

It is just as incorrect therefore to call the magnetic action in the wires electricity, because under certain conditions there is an exhibition of electricity at their poles, as it would be to call thunder clouds lightning, because under certain conditions they produce and yield lightning; or to call a tadpole a frog, because it becomes a frog; or to call an egg a chicken, because it produces a chicken; or indeed to call anything else by any name whatever, except its own.

Of course if the scientific world choose to reverse the names, and call magnetism, electricity, and *vice versa*, it would be all right, if the same name be applied only to the same property wherever it is found. But this is not done. A magnet by them is called a magnet, and its force is called magnetism. This force is exactly similar to the force in a telegraph line, and yet they call the latter electricity. This is what we object to, and wherein we desire to correct public opinion.

Magnetism therefore, and not electricity, is the great "imponderable agent" which governs and controls all the movements of the material world.

Instead therefore of magnetism being dependent upon electricity, the case is reversed, as will be shown in the next chapter, and all forms of "light" whatever, are consequently caused by magnetism under certain conditions.

We stated at the commencement of the chapter that all light was electricity. How do we prove this?

Electricity in its broadest sense, is combustion between the two classes of matter, wherever exhibited. We can find no combustion without the two classes, and we can find no electricity without combustion, in some form or another. Sunlight, lightning, gas, coal and wood fires, volcanoes, are all examples. As lightning is the best known form of electricity, we will explain how it is caused.

Gases are always rising from the earth, mineral as well as vegetable; these coming in contact in the atmosphere, by reciprocating, and reversing their poles, become opaque and form clouds. Thus combined they occupy less space than at first, and by the greater attractive power which is acquired by them, surrounding gases are drawn to them. This tends to produce wind, which still further increases the reciprocal action. The atoms when in form of clouds, being free, are reciprocally brought into immediate action, and according to their position, condition, and compounds, produce rain or snow, hail, fire or

Electricity, all those being the result of different unions of the same two classes of material, under varied conditions.

Where a cloud thus formed is insulated, and is moved within the influence of others, they resemble decomposing galvanic batteries, and act upon— or discharge into— each other, the lesser into the greater, by their instantaneous chemical action. The friction of the atoms produces combustion, and this is exhibited as electricity—being similar to that derived from the poles of a battery. When these thunder clouds with this excess of action, move within the attracting influence of an object connected with the earth, its mineral particles are concentrated into a ball-like mass of liquid fire, and through this point of contact—such as a lightning rod or church steeple—the earth, as the greater magnet, attracts the fire ball to its surface. Its contact, in passing, will destroy any small moist or large dry object, but will be immediately dissipated when coming in contact, either with the wet earth, or a large metallic surface. The remaining oxygen and hydrogen, or vegetable and mineral atoms, in the cloud, being shaken and brought together, then combine, form water, and fall as rain to the earth.

CHAPTER XV.

MAGNETISM.

Explanation chapter.—To show difference between Magnetism and Electricity.—Prof. Grove and Faraday.—Electricity not a force at all.—Arrangement of a galvanic battery.—How telegraphing is accomplished.—Telegraph worked by grass.—A few facts about magnetism.—Well known and not generally known.—Faraday's misfortune.—Born too soon.—The "Magnetic curves" explained.—Tyndall astray again.—Polarity of iron railings.—How the polarity of magnetism changes with position.—Sir Isaac Newton's apple.—The Law of gravitation upset.—How magnetism is a weight, and how it affects weight.—What Newton wished to discover.—The cause of deviation of compasses in iron ships.

This chapter is introduced more by way of explanation than anything else, in order to distinguish Magnetism from electricity. We described in our second chapter on matter and its force, the leading characteristics of magnetism, and every chapter since, has had more or less reference to it.

Being the universal force in nature, we could not explain the cause of any phenomena without mentioning it; and the consequence is, that nearly every succeeding chapter will also contain something new regarding it, according as the subject matter appears in different positions and conditions.

Prof. Grove in his " Correlation of Physical Forces," says :—
" Magnetism, as was proved by the important discovery of

Faraday, will produce electricity, but with this peculiarity—that in itself it is static; and therefore to produce a dynamical force, motion must be added to it. It is in fact directive, not motive, altering the direction of other forces, but not in strictness initiating them." If magnetism is powerful enough to alter the direction of, and to control other natural forces, we would argue that it is more powerful than any of them, and that in consequence, all others take their rise in it—the object of our whole work is to prove the truth of this theory.

Again he says:—"Magnetism can, through the medium of electricity, produce heat, light, chemical affinity and motion." But if it be acknowledged, as we have shown in the previous chapter, that the spark in a galvanic battery is combustion, caused by the force of magnetism in the ends or poles of the wires, *then electricity is not a force at all*, and can no more produce heat, light, chemical affinity, or motion, than it can produce, or have any control over itself. The very arrangement of a galvanic battery might show us, that electricity has nothing to do with the force in it; for if it had, the producing material would show it, which it does not.

Magnetism is an invisible influence, or force, which appears to have been observed only under particular conditions in iron, or steel, and a few other metals. But indications of a similar influence and action may be found in all substances, and in connection with all natural phenomena. It is traceable also in all atoms, and is thus found to be *an inherent property of matter.*

Let us examine the nature, influence, and action, of that force as observed in steel, and trace some of its effects to their source, then by comparing the same action in other substances, and forms of matter, the similarity and evidences of the same natural law will be observed in the whole of them.

Magnetism was first discovered in a peculiar kind of stone or metal, by its attracting pieces of iron; such stones were called magnets. By rubbing a magnet over a piece of steel, it was found to impart a force or power to the steel, so that it became a magnet, and when poised upon a point where it was free to act, its ends would incline towards the ends or poles of the earth. Magnets when thus poised, are observed to influence each other by attracting their opposite ends, or poles, and by repelling their similar ends.

At this point we would like to give an explanation of the "magnetic curves" formed by filings between two poles of a magnet, an experiment which has long been known to scientific men, but never explained. It was Faraday's misfortune, that he should have been born before the law of atomagnetism was discovered, for Tyndall in his New York lectures stated, that these magnetic curves, or "lines of force":—"so fascinated Faraday, that the greater portion of his intellectual life was devoted to pondering over them." It seems strange that we can explain in five minutes, what Faraday consumed a whole lifetime in only *trying* to discover, yet such is the case. These curves are caused simply by the attraction of the opposite poles of the magnet; and the reason they form lines, is that each hair, or branch of filings, must repel every other hair starting from the same pole, because similar poles repel, whereas only opposite poles attract. This is the whole mystery, and any one with a magnet and filings can speedily test the truth of it.

We have already said in our chapter on Heat, that the force of atomagnetism when undisturbed is in straight lines, and we see evidences of it everywhere. In the fibre of trees, in grass and rushes. In dissolving a piece of iron in acid, it is eaten away in lines. If the frost leaves on our windows are observed forming on a winter's morning, we may also perceive that the force is in straight lines, unless they are swayed either one way or other by the polarity of rival shoots. Thus we have seen two fronds starting from the bottom of a sash, a little apart from each other, and shooting out towards the centre, but on the points approaching they each repelled the other, which caused both to curve outward again.

On breaking a piece of shell ice, we have often seen on the under side, long lines of small spears of ice, formed as regularly the one behind the other, as a regiment on parade. Tyndall thinks that these magnetic curves will, by the progress of science, be found to represent a condition of the "luminiferous ether" which is "the mysterious substratum of all radiant action." We are only sorry for him that these curves will, in the progress of science, only prove that his theories of light and heat are all wrong.

To resume our facts regarding magnetism, particles of iron are attracted to both ends or poles of magnets, but not to their middle portion or centre. A steel magnet when bent to the shape of a U, or of a "horse shoe," shows its greatest force at either end, gradually diminishing towards the middle. This may be seen by placing the magnet in iron filings, when the ends by attracting the filings, are united, forming an arch of filings. A piece of soft iron similar in size to the ends of the magnet, if brought into contact with the ends, will be immediately attracted to the magnet. In this situation no filings will be attracted to the magnet, because the ends or poles of all the atoms are preoccupied one with the other. If the magnet be divided into minute particles, the like force and action will be found to prevail in each particle, or atom, and is merely diminished in proportion to the reduced size of the piece.

If a bar of steel remains fixed in a vertical position for any length of time, it will afterwards exhibit all the properties of a magnet, the upper end when placed in *any* position, will attract the "north" point of a compass needle, and the other end the south point. But if we take a piece of soft iron, every time its ends are vertically reversed, it is immediately changed in its magnetic polarity, to correspond with the attraction of the earth—that is, the lower end attracting the south point, and the upper end the north,—in these northern latitudes; in southern latitudes of course it is reversed. If we hold a pocket compass to the upper ends of any iron railings surrounding gardens or houses, or the upper end of a stove, or any fixture of iron whatever, we will find that the north point will be attracted, while the lower ends will attract the south point.

One of the strange scientific delusions of the day is, that if a piece of iron is struck several times with a hammer, it is converted into a magnet. It never seems to occur to those ingenious philosophers who perform the experiment, to try whether it is not a magnet without being hammered. From these facts, we infer that there exists an inherent force in the atoms of the iron, which must be under the dominating influence of a similar and greater force in the earth.

In the arrangement of metals for the operation of the telegraph, we again find that the force in the line and instruments, may be produced, diminished, or changed and controlled at pleasure, by the operator.

We will examine the arrangement, and observe the effects and causes of these changes.

To form a battery, the ends of two pieces of metal are placed in diluted acid, and by chemical action they are gradually dissolved. In this position they have no other magnetic force than the power of dissolving. Each piece has however two poles, the poles in the acid being similar, and those in the air being also alike. If the dry ends be brought into contact, the action of the ends in the solution is seen to increase; this is owing to the two pieces now forming one magnet with two poles, in place of four. The two poles lost at the point of contact, now merge their influence into the whole, and the poles in the solution become dissimilar—opposite poles (as explained before) then reciprocate, or attract one another, and thus we have the increase of the dissolving power exhibited, and an increase of force developed.

If the junction of the two metals be made by a long connecting wire, instead of close contact, the same action is continued between the poles or ends in solution. No more force is shown in the connecting wire, than in the middle of the magnet, because the force can by its nature, only be developed or exhibited at the ends or poles. If however we sever the wire or its connection at any part, the action is immediately checked at the ends, because they are now as before, two magnets and four poles; the poles in the solution being of course similar. Bring the separated ends slightly into contact, and the result is the electric spark; for as we have explained before, those ends act as opposing poles, and attract the mineral atoms towards each other in such a way, that they combine with the oxygen of the atmosphere, and *produce electricity and light.* By the same action in the solution between the poles, the metal dissolving unites with the oxygen of the water and throws off hydrogen, or forms other compounds, such as sulphate of zinc, etc.

It will be observed then, that a close connection of the wires throws the force into the ends that are in the solution, thus converting the whole line into one magnet; while a slight connection induces the poles of the two pieces, as two magnets, to reciprocate at their junction, thereby producing the resulting magnetic spark, known as electricity. Telegraph instruments, or "relays" as horse shoe magnets, forming poles, are arranged in the several offices along the line, and by them the operation of telegraphing is performed.

The magnetic force may be increased in the line to any extent, by multiplying and arranging a great number of pieces of any two kinds of metal, in a solution of acids, proportion being had to the quantity of surface metal, and the strength of the solution adapted to act chemically upon the metal; besides paying regard to the arrangement and amount of metal in connection.

Magnetism is latent (or active) in all atoms of matter, and may be brought into action, conditionally, in many ways.

The magnetic repelling force of atoms in dissolving substances, is equal to their attracting force in reforming new substances. The process of producing or reforming minerals, may be satisfactorily seen in the electro-plate, in the formation of the Lead tree, and in all kinds of crystallization. As all natural formations arise from the operation of this atomagnetic force, its action in dissolving and reforming minerals, vegetation and animals, upon and within the earth, is noticed in other chapters.

Electricity therefore no more flows along, or through the telegraph wire, than milk or water does; and if physicists would examine more into the nature and action of atomagnetism; not only in reference to metals, but in other departments of nature, they would have less failures in connection with telegraphing, and other similar operations, than have generally fallen to their lot.

As an incentive to new inquiry in this direction, it may be stated that we have seen a telegraph line charged and worked without the metal galvanic battery, by simply allowing the wire line, to come in contact with the long blades of grass growing in swampy ground. The force from the dissolving, or decomposition of the soil yielding the grass, affected the line in a similar manner to a mineral battery—and induced the magnetism in the blades of grass, on the occasion referred to, to supply the wire with a much greater force than was necessary for the ordinary working of the line.

In conclusion, the magnetism which is so abundantly possessed by iron, is also inherent in all other bodies, but in a much less marked degree. The magnet is only observed to exert its influence on iron, when the said iron is in a favourable condition for the purpose; so all bodies when in favourable positions, and conditions, principally attract or repel their own kind—although a large body, such as the earth, composed of every variety of substances, will attract materials of every kind.

The apple that Sir Isaac Newton saw fall by the law of gravitation, was attracted by the law of magnetism to the earth, from which it had been produced; because the earth contained the same kind of material as that which composed the apple. The tree kept the apple as long as it had the power to do so; but as soon as the attraction of the earth became more powerful than the tenacity or attraction of the tree, it fell. Sir Isaac Newton's grand discovery about the law of gravitation, may then merge into, or become a branch of, the more comprehensive and *universal* law of ATOMAGNETISM, for by the following simple experiment, we prove the theory of gravitation to be *not universal*. Take a bar of steel, not magnetised, and balance it in the middle, then merely touch it with a powerful magnet, to charge it with magnetism, and one end,—that which attracts the south point of the compass—immediately falls. Can gravitation explain this? No. But atomagnetism can, for the earth being also a magnet, it attracts that pole of any other magnet, which is the reverse of its own polarity. We at this part of the earth—north of the equator—are placed far towards the north pole, consequently the opposite or south pole of a magnet is attracted to the earth. If the same bar is balanced anew, and its whole polarity is again reversed, the other end will fall. The experiment may be repeated again and again with the same result.

Sir Isaac Newton himself was evidently not satisfied with his discoveries, for he said:—"To devise two or three general principles of motion from phenomena, and afterwards to tell us how the properties and actions of all corporeal things follow from those manifest principles, would be a great step in philosophy." This we believe atomagnetism can do.

Before closing this chapter, we would like to say a few words, on the deviation of the compass in ships, for the loss of life and property on the North American coast and other shores of late, has, through this and other causes, been lamentable.

Notwithstanding the long time that the mariner's compass has been used, scientific, as well as practical men, seem to be as ignorant as ever of the influences to which the needle is subject, and to have learned absolutely nothing in this respect, to efficiently protect their vessels from danger. Since the introduction of iron vessels also, accidents have become more numerous, even as high, compared with wooden vessels, as 8 to 1, and yet the only invention brought into use to prevent the deviations, is one of the most stupid, not to say dangerous, that could have been thought of, viz. : the placing of fixed magnets on the deck of a vessel under the compass. For the one thing absolutely required in the use of a compass, is perfect freedom from all magnetic influence, except that of the earth's polarity, and yet fixed magnets are placed in such a manner, that if the ship alters its position, or its course, by storm or otherwise, the true marking of the compass is lost altogether.

In wooden vessels there is not so much danger of deviation as in an iron vessel, and the only care necessary, is that no iron of any description be placed near the compass. Any metal required for the fitting up of the ship should be of composition. Care also should be taken with the description of cargo, stowed under or near the compass.

We read lately of a Captain who sailed from New York for Japan, with a cargo of petroleum in tin cases, and after being out a day, found he had been going more southwardly than he expected. On making an examination of the compasses, and trying them in different parts of his ship, he found that anywhere within three feet of the main deck they were alike, but on raising them six feet and over, they showed differently. He thus discovered that the tin cans had caused the deviation, as they were merely sheet iron covered with tin; so during the remainder of the voyage he steered with a compass fixed half way up the mizen mast. He stated that if he had sailed from Boston, in thick weather, instead of from New York, it was highly probable he should have lost his vessel on Cape Cod. But it would not require a full cargo of tin cans to make such difference in the compass, as only one tin can, if sufficiently near, would affect it. Before iron ships came into use, we heard an account of a difficulty from compasses on board one of H. M. Ships, the officers finding an error in her position after every night's sailing. An investigation showed it to be caused by the officer of the night watch, carelessly placing his speaking trumpet in the binnacle, alongside of the compass, when he came on deck.

The influences to which compasses in iron ships, or steamers, are subject, are more numerous.

The first great cause of deviation, is from the iron sides and projections of the vessel, all iron projections, such as stanchions, davits, etc., acting as separate poles to the ship's magnetism.

The uppermost sides of the ship and the projections, will always—in northern latitudes—attract the "north" point of the compass. Thus if the ship rolls heavily from side to side, the compasses will move from side to side with every roll. But if the ship sails with a steady list to one side, the north point of the compass will keep inclining to the uppermost side as long as it is so. A vessel from this cause is apt to go a long way out of her course; and instead of attributing the trouble to the deviation of the compass, currents are stated to have caused it.

The steering apparatus of many steamers in the wheel house, is also composed of iron and steel, and these act as the poles of the ship's magnetism. They not only at all times affect the compass, but the effect varies, as the helm is shifted hard a-port or starboard; the joints of the screw steering gear sometimes advancing towards, or receding from, the compass, thus attracting the north point of the needle, or repelling the south. Again, when steam is up in the boilers, the magnetic influence of the whole machinery is increased by the action from them. Care in observations should also be taken in approaching a rocky iron-bound coast, particularly such as that of Nova Scotia, where the S. S. *Atlantic* and the *City of Washington* were lost, because the strata of rock is nearly vertical, which in consequence allows the escape of mineral emanations from the interior of the earth, to influence the ship's magnetism, and thus to alter her compasses. A considerable deviation might thereby be induced without any apparent cause for it.

The best safeguard, with due care, would in most instances, be the fixing of a compass at such a distance above the deck, as would be beyond any local influence. And yet with this precaution, the most absurd mistakes are made, for in many steamers we have seen the compass elevated on the mast, but fixed there under *iron cross trees*. They are thus led into the very danger, which, in elevating the compass, they expected to avoid.

For the safety of life and property, all sailors in particular, should know these simple facts; and all iron vessels should be examined and reported upon, concerning such parts of their construction or fittings, as are liable to derange the compasses.

CHAPTER XVI.

SOUND.

Difficult problem in Science.—Prof. Tyndall's explanation not satisfactory.—Sound vibrations and light vibrations.—Sound generates heat.—How long fifty organs would take to heat St. Paul's Cathedral? Sound in summer and winter.—How we hear fifty sounds at the same time.—Echoes.—New theory of Sound.—A sympathy between the mineral atoms of matter.—Iron a better conductor than wood.—If a man has sympathy why should not an atom?—Dancing flames.—Tyndall's new theory of Sound.—Experiments at the South Foreland, England.—Vapour in layers.

An explanation of the phenomenon of Sound we consider to be one of the most difficult problems in science, and the manner in which it has been explained by Prof. Tyndall, and others, is far from satisfactory. For instance he says, sound, light and heat, are all caused by the vibrations of the atoms of the atmosphere. But sound, light and heat, all travel at different rates of speed, and in order to surmount this difficulty he says :—" they all vibrate different ways," a most empirical and yet safe assertion, for it is beyond the power of any experimenter, or microscopist to demonstrate how atoms vibrate.

Again he says :—" Sound generates heat." " Every sonorous vibration which speeds through the air of this room and wastes itself upon the walls, seats and cushions, is converted into the form with which the cycle of actions commenced :—namely into heat." Theories like these are very easily advanced, but rather difficult to prove. We would like to ask the professor how long under the most favourable circumstances for accumulation, it would take fifty ordinary church organs to heat St. Paul's Cathedral?

The experiments which are made in every Scientific Institution, to illustrate the various phases of sound and vibration are both numerous and beautiful, but we fail to see that sound is vibration of the atoms of the atmosphere only, and that it cannot exist independently of such motion.

For instance, if we sound a bugle on a warm summer's day, and again on a clear frosty day, it is heard with twice the distinctness, and at twice the distance, on the latter occasion, although the vibrations are theoretically the same.

Again metal is the best conductor of sound, and a long piece of iron, from its mineral character, will yield more sound when struck, than a similar piece of wood, although in theory, the vibration is necessarily the same.

Suppose a bar of iron, twenty feet long, by twelve inches square, lay on the ground, and we strike it with a tiny hammer, sound would result, although it is impossible that the bar could vibrate. Again if the bar were suspended and struck with the same force, we would have more sound, and yet there would be no vibration, thus showing that the *position* of atoms has a great deal to do with sound. If the material composing a bell were cast in any other shape, it would not yield nearly so much sound as before the change.

While we admit that in many instances, vibration causes sound, and particular vibrations cause particular notes, yet we could no more say vibration of itself was sound, than that a hammer was sound—for if the vibration of a bell caused sound, then the hammer caused vibration—these are only secondary causes; *sound must be something deeper, some innate property of the atoms entirely independent of vibration or motion.*

Vibration also will produce sound from a tuning fork, or a violin string, but that a special vibration of the air for each particular note, speeds from the instrument through the atmosphere to strike the listeners ear, we deny. If this were the case, how is it possible for us to hear fifty different sounds at the same time?

It may be said that echoes prove there must be vibrations in the atmosphere, for the *sound waves* strike the obstruction, and are forced back again. We disbelieve in sound waves altogether, and think it evident that if there were such phenomena, in striking the obstruction, the waves would be so changed, that a note would be sent back of a different character from the one first sent.

Taking everything into consideration, we submit the following theory of sound :—*Sound is a property of* SYMPATHY *between the mineral atoms of matter, induced in the first place by friction or vibration.*

Thus we hear better on a frosty day, because the atmosphere has more mineral atoms in it than on a summer's day. On the like principle, iron must be a better conductor of sound than wood.

Thus also we can hear any number of sounds at the same time, for the sympathy of the atoms naturally repeats them all, whereas the vibrations of the atmosphere, as defined by the theory referred to, would neutralize one another.

So also in echoes, the sound is accumulated at the point of obstruction, and must necessarily come back without change.

Human beings are endowed with a great amount of sympathy and thus will naturally laugh, or cry, or dance, just as the sounds they hear impel them. If human beings, who are only made up of a conglomeration of atoms, should have sympathy, is it not likely that each individual atom is also proportionally possessed of it? If a man should feel impelled to dance while under the influence of music, why should not the sympathetic atoms in a flame cause it to dance also?

There are many of the experiments relating to sound not easily to be explained, and many theories scarcely admitting of demonstration, but we think that fewer difficulties will present themselves by the explanation we have given, than by any other.

Prof. Tyndall lately delivered a lecture at the Royal Institution, on Sound, giving an account of numerous experiments made at the South Foreland, England, with steam whistles, trumpets and cannons, in order to determine the distance at which sounds could be heard at sea. The result of these observations is a New Theory of Sound, and it just tends to show how much dependence is to be placed on any theory of abstruse science, where the *imagination* is allowed considerable latitude—no permanent basis of natural science being established, which would enable any one to prove or disprove any startling assertion.

The theory is not complete, but it is to the effect, that the imagination has to picture vapour from sea and land, rising in layers; these layers presenting "reflecting surfaces" to the passage of sound :—" In the relative homogeneity of the atmosphere, or its being split up into many layers, we have a clue, which may enable us to arrive at a knowledge why sounds of equal intensity, will travel further in some days than others." Long discussions and lectures will probably be the result of this discovery, till, when it is on the point of being universally adopted, another clue will be unfortunately discovered.

CHAPTER XVII.

WATER AND RAIN.

Fire not so powerful as Water.—Water in granite.—Herschell on Rain.—Rain caused by chemical action in the atmosphere.—The Rain guage.—Rain forms in the lower atmosphere.—Proctor and Kamtz on the reason why.—Rain shot out from clouds.—Herschell on Rain storms.—Climate of North America changing.—Egypt cultivating the Palm for Rain.—Forests and vegetation cause Rain.—Herschell's reason why, a failure.—Drainage said to be bad.—Chicago, St. Louis, once unhealthy.—Why.—No large city unhealthy.—No air in Water.—Fishes gills used for filtering food, not for breathing.—The air they need produced from digestion.—Can we produce or bring down Rain?—Great battles in America were followed by Rain.—The cause.—Conclusion.

THE subject of the present chapter, deals with one of the most powerful agencies in nature, by the medium of which, all formations,—animal, vegetable, and mineral,—are by turns produced, dissolved and again reformed.

Many scientific men assert that Fire is a more powerful agent than Water. For dissipating and dissolving, it is so, but it does not rebuild, and cannot re-produce. It has been generally considered that all minerals, diamonds, and precious stones, have been formed from the action of fire, also granites and coal; but scientific opinion is gradually veering round to a belief in the more powerful agency of water.

Strange to say, granite when examined by a microscope, is found, by Dr. Sorby, to have minute cells filled with water, a most emphatic demonstration that it has been formed by the medium of water; yet this has been explained away on the ground that it is condensed steam, or vapour, which was present during its formation. How steam could exist where every substance (as believed) was in a molten state, and confined to the interior of the earth, is more than any one could imagine possible.

If we except the atmosphere, there is nothing so abundant on this earth as water; for the oceans are larger than the continents, and the land even abounds in lakes, and rivers, while the atmosphere itself is continually pouring down deluges of rain. What is this wonderful element, and how is it produced?

Natural Philosophers tell us it *is* composed of oxygen and hydrogen—eight parts of oxygen, to one of hydrogen—and that with these gases "artificial" water can be easily manufactured in a laboratory. They also tell us that the most powerful combustion, is produced by a combination of the same gases in different proportions.

Agreeably with our atomagnetic system, we prefer to use simpler language and say :—Water is the result of the simplest combination, next to air, of the two classes of matter in the form of gases. Moreover, because these gases can be made on a small scale only by a certain chemical process, our teachers would have us believe, that they are not to be found naturally in sufficient abundance in the atmosphere; and consequently, that rain is not formed by their combination in it. But they forget, or rather they do not perceive, that the earth itself is a vast laboratory, where hundreds of nature's gases, of every description, are continually being manufactured by the agency of water in the interior, and thrown into the atmosphere; where they again reform and descend in the shape of rain, or snow, or hail, to repeat the same transformation below.

They account for rain, therefore, by saying that all the water that comes down to us in that form, must have arisen first as vapour, and so remained for a time in invisible particles in the atmosphere, till it accumulated and fell again. Thus Prof. Tyndall in his "Forms of Water," says:—"Solar heat is the true origin of Glaciers. The sun acting on the ocean within the tropics, causes an exhalation which floats away as clouds to the Polar regions, as well as the high mountain ranges, where, in each case, the clouds yield up their contents as snow or rain." The too common practice of tracing everything to the sun, is something to be deprecated, and just as absurd, as to be for ever blaming Adam for all the ills and miseries that afflict the human race. There is no necessity for searching for a remote ancestry for any natural phenomenon, when its own immediate cause is explained. Apart from this, the theory is incorrect, as we will show further on. Thus also Sir John Herschell in "Good Words," 1864, says:—"Common sense assures us, that all the rain, etc., which falls from the skies must have originated in the sea, and must (if the present state of things is to endure) find its way back to it." Common sense is a very good guide for a man's actions, but a poor guide to the study of science, unless the given principles of science are correct. In our opinion, this theory would be merely distillation, which is an induced process, whereby the particles of water are expanded by heat, when confined in a vessel, so that their properties or compounds, are in no way altered by coming in contact with other compounds or gases; and, if cold be applied, they then return to their former condition as water. Evaporation, on the other hand, is a chemical action—induced by *cold* as well as by heat, and is a throwing off into gases of the particles of the material acted upon. Thus we have seen the wet muddy streets of a city in spring, dried up in a few hours by a piercing north wind, and such clouds of dust raised, that would have taken the sun a day or two in summer to have accomplished.

When bodies are evaporated into their original elements, either by heat or cold, or fire, or water, and then allowed to come into contact with other gases in the atmosphere, those compounds will be chemically changed into a variety of new compounds, and are thus in a position to produce a variety of atmospheric phenomena, such as rain, hail, snow, fog, clouds, lightning, thunder, auroras, etc., etc.

If all bodies can be converted into gases or their original elements, water cannot be an exception. The main scientific objection to this, is the statement given by professors of chemistry, that oxygen is the most universal gas, for it is found in connection with every other gas on earth, while hydrogen is only found beside metals. This is only apparently so. Vegetable gas must, as a matter of course, be contiguous to vegetation. It is also more dense than mineral gas, consequently we find it close to the earth. But as we ascend into the atmosphere, the air becomes more cold and rarified, so that we breathe with difficulty. This atmosphere cannot be composed principally of oxygen or vegetable gas, for no vegetation grows on the tops of high mountains; it must then partake more of the hydrogen or mineral gas. If a large quantity of this oxygen, should come in contact with the hydrogen, the result would probably be cloud and then rain, not on account of the temperature of the gases, but of their opposite characters causing a reciprocation and chemical action. Let us give a common illustration. The weather (in winter) has been cold for several days, with the wind blowing from the north, so that everything is frozen hard. Suddenly the wind veers round to the south. The consequence is we first have snow, (which is a compound of the gases when they are both cold or below 32° F.) and as the south wind prevails, the temperature of both gases rises, or rather the south wind so overpowers the north that it turns to rain.

What is the philosophy of this?

The north wind is composed of a large proportion of mineral atoms—emanating from the pole—which permeates everything, so that when the south wind blows with its large allowance of vegetable gas, brought from the region of the tropics, or where there is a warm sun and abundance of vegetation, there is an immediate change of combination with the result mentioned.

The rain or snow which deluges us, does not necessarily come from the Gulf Stream, or result from the evaporation of the Atlantic, as is generally assumed, but from the atmosphere which immediately surrounds us. Rain also does not necessarily fall from a distance of some hundreds of yards above us, but may be formed only a few feet over our heads. This is proved by observations which have been taken lately, showing that the nearer the rain guage is kept to the surface of the earth, the greater is the rain-fall indicated. The process of rain formation is everywhere in the atmosphere up to a certain height, and even down to, and on, the surface of the earth.

Thus it is we have fogs from a similar union of gases. The opacity is occasioned, we believe, from the poles of the atoms

—by reciprocating with others—being all turned topsy turvy. When the atoms have combined and are in position, they are again transparent. Thus if we allow sugar to crystalize quietly it will be transparent, but stir the syrup, and it becomes opaque. If water freezes on a calm evening, it is transparent, because the atoms are undisturbed,—thus we have seen the fish in the East River of Pictou, Nova Scotia, gliding along under the ice, and yet the latter was so thick that horses and sleighs were travelling over it—but if the wind blows during the process of freezing,—as we once observed on Dunsappie Loch, near Edinburgh, Scotland,—the ice, no matter how thin, becomes a milky white and obscure. So it is with many other solutions.

Mr. Proctor, in an article on Rain in the Intellectual Observer, alludes to these observations with the rain guage, and says Kamtz, explains it thus :—"A drop carries with it the temperature of the upper region of air, and condenses on its surface the aqueous vapour present throughout the lower strata of the atmosphere, as a decanter of cold water does when brought out of a room."

But the mere immaterial sensation of cold, could not produce a material substance like rain. The *cold* must either be a material substance itself, or it must have a material to represent it. We say cold, as experienced by us, is caused by hydrogen or mineral gas, being a property of it. If then a drop fell from the upper atmosphere and enlarged on entering the lower, it would not do so merely from the temperature, but because the hydrogen in the drop combined with the oxygen it came in contact with. Water forms on the outside of the decanter in a similar manner, not by reason of the temperature, but by the proximate cause of the temperature, the ice, sending forth mineral gas, which combining with the oxygen (or animal and vegetable gas in the room) forms water.

Kamtz's theory, such as it is, is supported by several well known men, but Sir John Herschell applies the Dynamical Theory to it, and because it does not stand the test, he opposes it, but offers no explanation in return.

Mr. Proctor, in concluding his article, thinks there is no difficulty in explaining the phenomenon, for he has observed that when rain is falling heavily, small specks are seen flitting among them, which could not be caused by the collision of the drops, for they all fall parallel. This, he says, shows rain is generated in the lower as well as the upper strata of atmosphere; for these specks are observed to be water. Why this is so, however, he does not profess to explain. A simple acquaintance with atomagnetic law will settle the question for ever. The idea that rain drops all fall parallel is not strictly correct, for we have seen them shot out from a cloud (like rays of sunlight), thus indicating that there is a force in the cloud which ejects, as well as forms the rain, when the combination of the gases or materials present, become repellent to it.

The theory of rain storms is very vague at the present time. Sir John Herschell, in the article we quoted from previously, asks:—"Is it in any degree in the power of man to alter the weather? This he tells us is not so absurd a question as it may appear; for from the registers of rain falls which are kept all over England, the rain would seem to have a preference for some places more than others. All well wooded districts are observed to have a large yearly rain fall, while waste barren moors are dry in comparison. He also says:—"The rain fall over large regions of North America is said to be gradually diminishing, and the climate otherwise altering in consequence of the clearance of the forests; while on the other hand, under the beneficial influence of a largely increased cultivation of the palm in Egypt, rain is annually becoming more frequent. Lakes are cited in what was formerly Spanish America (that of Nicaragua, if we mistake not, is one) whose water supply (derived, of course, from atmospheric sources,) had been so largely diminished, owing to the denudation of the country under the Spanish Regime, as to contract their areas, and leave large tracts of their shores dry; which, now that the vegetation is again restored, are once more covered by their waters."

The reason why trees attract the rain is, Sir John says:— "The foliage of the trees defends the soil beneath and around them from the sun's direct rays, and disperses their heat in the air, to be carried away by winds, and thus prevents the ground from being heated in the summer; while on the other hand, a heated surface soil, reacts by its radiation on the clouds as they pass over it, and thus prevents many a refreshing shower which they would otherwise deposit, or disperses them altogether."

We confess we cannot understand, neither does he explain, why a soil that is not heated will attract rain, or "vice versa." How much simpler would it not be to say, that well wooded districts attract or produce rain, because there is always an excess of vegetable gases around them, which are thus always ready to combine with any mineral winds that may be blown over them. While on a barren moor which has no sufficient supply of oxygen, reciprocation seldom takes place.

"Drainage"—Herschell also says, "is bad, for it cuts off a

great deal of the supply of local evaporation, which is the material element in the amount of rain fall."

We admit that drainage lessens the rain fall, not for his reason, but because there is less concentrated activity in vegetable growth, and consequently a smaller supply of vegetable exhalations necessary for forming rain. Still we would not say that drainage is bad, for a clump of trees at the end of a field, would counteract the effect of drainage over a hundred acres.

These facts regarding vegetable exhalations, give us an insight into the cause of fever and malaria. A hundred years ago the sites of many of those great western cities in America— St. Louis, Chicago, Cincinnati, and others, were fever swamps, full of all kinds of malaria, striking down the strongest man that took up his abode in their midst; yet they are now as healthy as any of the cities of the Republic. Why were these cities unhealthy? Because there was too much vegetation present with decomposing vegetable matter, and in consequence more oxygen or vegetable gas in the atmosphere, than is suitable for the health of man. We never hear of a city being a place of fever and ague, especially one which is largely built of brick and stone, unless the drainage and cleanliness are defective, for the vegetable emanations in its area are very scant. But allow the cities to go to ruin, let them be overrun with vegetation, as in the case of Palenque and Uxmal in Central America, or those magnificent ruined cities of Cambodia in Siam; and they become as dangerous to the life of man, as the worst ague swamps known.

A curious fallacy expounded regarding water, is, that it contains a quantity of air. Prof. Tyndall proves this by saying that bubbles rise to the surface when water is boiled. But if we boil the water long enough, it will all vanish into air, or its original elements, therefore according to Tyndall's own showing, water is all air. If water contained air, we should think it would be enclosed when frozen into ice. But Prof. Tyndall declares that it is not so, "for although ice is full of small bubbles, they are not filled with air."

Mr. H. Higgins in "Fraser's Magazine" 1870, on "The water we should not drink," says:—"A considerable volume of air is absorbed by water." "In this and in other ways (bubbling over falls and among rocks) water receives atmospheric air, without which it would fail of the purposes for which it was ordained. It is necessary to the existence of the creatures who live in the water, and for the continued purity of the water itself."

This statement is plainly incorrect, for if water were capable of absorbing air by a process of agitation, then when we introduce a small quantity into a large vessel of water, it ought to permeate itself through it, and not show itself; but it invariably (except when confined by pressure) comes to the surface in bubbles, and disappears, thereby showing that water has an aversion to air. Water may be, however, saturated with air by mechanical pressure, as in soda water or champagne, the pressure while corked not allowing the air to escape.

The idea that fish use their gills to assimilate air for themselves, is a grand mistake; for the motion of the mouth and gills of the fish, is not a breathing process, but one for filtering, or separating their food from the water with which it is combined. This can be proved by observing the peculiar construction of the gills of certain kinds of fish, and the nature of their food. The shark and the dog-fish have but little or no gills, (neither has the lobster) because their food is solid, and no filtering is necessary. But the whale, the herring, the mackerel, and others that are provided with extensive gills, take in a small description of animalculæ for food with the water; and having discharged the latter through their gills, as we see the whales in particular do, the food is then left in a state for them to swallow. From the fact already noticed that water does not contain any fixed air, the air necessary for the support of the fish is produced from the dissolving of the food in the stomach, during the process of solution and digestion.

The idea that air is generated by the dissolving and decomposition of food in the fish, is supported by deduction from natural facts, for we find that *all* decomposing or chemical action generates a gas or air. The drowned body of an animal, for instance, becomes inflated with gas—not from air contained in the water—but from the process of decomposition by the water—and the body rises to the surface, where it floats.

The question has often been asked, can we call down rain at will? and it has been answered in different ways. We answer that under certain conditions it is possible, but the expense of doing so, would be greater than the value returned. During the late war in the United States, it was observed that every great battle fought in the South was followed by deluges of rain, and violent wind. It was then stated by many, in consequence, that we could easily bring down rain by merely discharging cannon. But during the Franco-Prussian war, the like phenomenon was not observed, so the idea was declared a myth. Since the war, some scientific American wished his government to lend him a few hundred guns to settle the question; but the government very properly refused. It was thought the concussion of the guns caused the rain, but no vibration however great will cause rain to fall, unless the necessary materials are present in the atmosphere to furnish it. The rain was in part produced from the material used in firing the cannon, viz.: the gunpowder. As powder (sulphur, saltpetre and charcoal) is composed principally of mineral ingredients, these in passing into gases consequent on explosion, reciprocated with the abundant vegetable gases that filled the atmosphere in the Southern States, and in the first place formed clouds, then wind and rain. The same result would have followed, if the powder had been merely set on fire without the use of guns. The reason why no rain appeared on the occasion of the Prussian battles was, that they occurred in winter, and in places where no excess of vegetable gases existed in the atmosphere.

In conclusion, as animal and vegetable bodies are composed of the same material elements as water, let us apply our knowledge to a practical purpose. We cannot then perform our ablutions too often, and the more frequently we wash ourselves without becoming altogether amphibious, the more healthy we shall become. We cannot also, if at proper intervals, drink too much water, but it should always be preferred of the same temperature as the body. The frequent use of iced water is often injurious to health, and there is no doubt that in the present day it is used much too frequently.

CHAPTER XVIII.

DEW.

Chambers's Journal.—Baptista Porta nearly discovered the true theory of dew.—Thought dew was condensed from air.—Aristotle thought it was condensed from vapour.—Muschenbrook kept back Meteorology one hundred years.—Great discoveries often foiled by the stupidity of the world.—Dr. Wells said to be the discoverer of the true dew theory.—The radiation of heat, the basis.—The cause of moonblindness.—Dew forms most readily on vegetation.—Arguments against radiation.—Observations with wool packs.—Position everything.—Calm and clear evenings essential.—Dew is water.—Produced in a similar way.—The cause of fog and hoar frost.—Hoar frost spears of ice.

In "Chambers's Journal" for 1868, is an article on the above subject, which pretends to settle the question definitely about the formation and phenomenon of dew; but, like most other explanations of natural phenomena, it fails for want of a correct apprehension of natural law.

We find in it, that one Baptista Porta, nearly discovered, (according to our view of it) the true theory of dew; but his notions are sneered at altogether, because they do not agree with dynamical principles.

The old idea was, that dew was precipitated from the moon and stars; which also shed down cold. But the writer of the above mentioned article states, that so far from shedding cold on the earth, astronomers and physicists show, that an important portion of the earth's heat supply is derived from the stars. A statement just as absurd as the other.

Porta denied that the moon and stars had anything to do with dew. He discovered that dew was sometimes deposited on the inside of glass panes; and again, that a glass bell placed over a plant in cold weather, was more copiously covered with dew within than without. He thought his observations justified him in looking on dew as condensed—not from vapour as Aristotle thought, and as is now believed by the scientific world generally—*but from the air itself*. This, although not entirely correct, was a remarkable discovery, and had it been believed in then and worked up to, a great deal of blundering might have been avoided, and our knowledge of meteorology would have been much further advanced than it is.

Dew was generally supposed to *fall*, and people still continue to speak of its falling, but Porta's experiment showed that it rose from the earth, that it was an exhalation from the ground and from plants. In making observations to establish this view, Muschenbrook found that dew forms much more readily on some substances than others. This was supposed to be damaging to Porta's theory, for dew neither seemed to fall, nor to rise, but to be caused in a great measure by the nature of the substance on which it was found deposited. Had this circumstance only been searched into more minutely, it would have shown still more conclusively that Porta was right, and greater discoveries might have resulted from it. Yet it is always the way of the world, we are often on the eve of wonderful discoveries, by great minds who are ahead of their day and generation, but which are foiled by the stupidity of those for whose benefit they are designed, and who have not the brains to understand their tendency.

The true theory, we are told, was at length discovered by Dr. Wells, of London, who made a series of observations with a number of little *wool packs*, and the result was: — "the rate of the deposition of dew, depends upon the rate at which bodies part with their heat by radiation. If the process of radiation is checked, dew is less copiously deposited, and vice versa." For instance, we are told that the earth is continually throwing off its internal heat. If there are any clouds, the radiation is checked at night and there is no dew, but when it is clear, plenty of dew is the result. This, however, is incorrect, for we have often observed that on a clear night no dew is formed, because there was wind. A calm night is just as essential as a clear one.

Dr. Wells also asserts moon-blindness to be caused by the want of clouds to check the radiation of heat from the eye, which consequently becomes chilled. Moon-blindness, therefore, according to his theory, ought to be as frequent on clear starry nights, as on clear moonlight nights, yet we never hear of persons being afflicted by it. This assertion, consequently, is just as much to be relied on as the others.

Let us now endeavour to find out the origin of dew from atomagnetic law.

Some objects have more dew formed on them than others. Grass and bushes in the morning are found covered with dew, while rocks and gravel roads are perfectly dry. Metal we are told radiates very little heat, and no dew forms on it, yet if we place a piece of metal among grass, it will be covered with dew; while a piece placed on the road bed, will have none. What is the reason of this? It cannot be that it radiated its heat quicker in the one place than in the other.

This is one argument against the radiation theory. Again, if dew be formed by radiation, why is it that a glass bell placed over plants is covered with dew inside, but not outside, and that there is no dew on the plants themselves. Obstructions between an object and the sky, we are told, check radiation, and prevent the formation of dew, *yet here we have the plants radiating, and dew forming on the obstruction!*

The whole series of observations, which led Dr. Wells to advance his theory, seem to have been conducted in a loose manner. He had a number of little wool packs which he exposed at night, some he covered, and some he did not; and it was by weighing the amount of dew which each contained, that he became convinced that he had something to work on. He found, generally, that those packs which were hid from the sky, contained more moisture than the others. This ought to have been contrary to his theory, but he does not seem to have thought so. It does not appear that he thought the situation of his wool packs of any consequence, and if so, his observations are of no value whatever, because *position* had every thing to do with the deposition of dew on them. If he placed them in the centre of a broad road, with little vegetation around it, no dew would ever reach them; while if he placed them over grass, or among trees, they would be heavy with it.

It will be observed that the action and formation of dew, has thus been explained by secondary causes only. We are told, for instance, that glass radiates heat better than metal. Why it does so they cannot show us. That it radiates heat at all, is an assertion which is dogmatically laid down, and cannot be proved, (heat being merely a property or condition of matter.)

Two facts remain, however, that with few exceptions, it is only on calm and clear evenings that dew is formed, and the warmer the day has been, the greater is the amount of deposition. It is evident, therefore, that some part of the material necessary is derived from the upper atmosphere; and as dew is found principally beside vegetation, the other material comes from it.

What then is dew? *Dew is water, and it is formed in a similar manner, by a reciprocal or chemical action between the two classes of matter, viz: the vegetable emanations or gases rising from plants, and reciprocating with that from the upper mineral atmosphere, which on a calm evening descends and meets it.*

If the sky is cloudy, the same action is going on in forming the cloud, and making the cloud larger, therefore it cannot work on the earth's surface; and if it is windy, the gases take of necessity a horizontal direction, and do not meet in a suitable position to form dew, for each class is blown onwards at its own level, forming cloud.

Sometimes the differing gases will be more dense, and meet in the lower atmosphere, the result is a fog; and if the cold mineral air is more powerful, the result is hoar frost. On the latter occasion, if a rounded stone or stick be examined, it will be observed that the spears of ice are longest at the highest points of the surface, and that they diminish on the sides, till they degenerate into a mere glistening powdered dust. Thus showing, in a humble way, the direct continuity in the action of atoms, before spoken of, and that the formation is caused by a direct reciprocation between the ascending, and the descending gases.

CHAPTER XIX.

THE ATMOSPHERE AND STORMS.

Atmosphere said to be composed of oxygen and nitrogen.—An impossibility.—Air in no two places the same.—Balloon explorations.—Guy Lussac.—Everything with life has an atmosphere.—The atmosphere of the African.—Impossible to get rid of it.—The earth a living body.—Has an Atmosphere composed of its own materials.—The Atmosphere composed of hundreds of different compounds of materials.—STORMS: Sir John Herschell and Prof. Rogers on Storms.—Magnetic curves from the poles of the earth, the cause of wind and storms.—Cause of Equatorial Calms.—Maury on cyclones.—Description of a so-called Circular Storm.—Hints for Weather Prophets.

IN school books we are told that air is composed of "20 parts of oxygen to 80 parts of nitrogen." What nitrogen is we are not told, but from the way it is made artificially, we should imagine it to be of a vegetable character. As oxygen is of the same nature, there must be some of the other elements—as a mineral—to counteract and modify the vegetable, else the atmosphere would be unfit for man or beast to live in. Besides, we do not find the air in any two places to be the same. The air of the city is different from that of the country, and the sea air, from the air near a lake. The air at the tropics, is different from that of temperate or polar regions, and the air on a mountain top, from that in the valley. All the atmospheric phenomena we see and hear, fog, rain, thunder, lightning, hail, snow and cloud formations, show us that the constitution of the atmosphere is continually changing. Therefore, to say the atmosphere is composed of two kinds of gases, in certain proportions as elements, and nothing else, is simply unscientific.

Researches have been made in balloons, to discover whether the atmosphere decreases in density, and temperature, at any regular rate, but with no very successful results. Guy Lussac, a Frenchman, found the temperature at 22,000 feet high, to be 15°, while other balloonists found it 30°, below zero; thus showing that the temperature and composition of the higher atmosphere, varies in a like degree to that on the earth, and is subject to the different conditions and positions of its surroundings, or the material with which it comes in contact.

What then is our atmosphere composed of? We have stated before, that every thing with life, or having a life action, has an atmosphere, consisting of emanations from itself. In life action, the plant or animal is continually throwing off waste material into the atmosphere, through its outer covering. This keeps around it an atmosphere of its own, peculiarities of which may generally be detected by the organ of smell. Thus we have the atmosphere of a rose, sweet brier, a horse, a cow, a rabbit, an African, or a European. The atmosphere is thus part and parcel of the body, and it would be as impossible for animals to avoid, or cut off their shadows, as their emanations. The earth then is a vast body, having a life action. It is composed probably of hundreds of different kinds of mineral and vegetable atoms. These, by the action of water in the interior, are continually forming, dissolving, and reforming, into the different substances found in nature's arcana, and all the time throwing off gases, which find their way through fissures in the rock, through volcanoes, and in other ways to the surface of the earth; where they again take up a position in the atmosphere according to their density. The light mineral gases all ascend and take up a position in the higher atmosphere, while the vegetable gases being more dense, remain near the surface.

Just also as a man, or other animal, has a coating of hair on the skin, so is our earth bristled all over (except at high polar latitudes) with an edging of vegetation, which is thrown out on every side by the atomagnetic repellent force of nature, after being produced by the dissolving process of water, and elements, on the surface. Thus thousands of varieties of trees and plants have been, and are being, continuously produced, each with a different exhalation; because composed in varying proportions of different vegetable and mineral materials; consequently thousands of differing vegetable, as well as mineral, gases, mingle in our atmosphere. It will be seen, therefore, that to define our atmosphere as composed only of two particular gases, is sheer nonsense. Enough it is for us to know, that as we ourselves and every living thing, are composed elementally of certain proportions of mineral and vegetable atoms, so the atmosphere adapted for our use is also composed of certain proportions of the same: and, as in overloading the stomach with one description of either vegetable or mineral food, we bring on disease and death, so wherever there is an atmosphere with an excess of one gas or the other, it is also a dangerous place for us to live in. We should therefore learn the nature, and actions, of those materials, and be able by what we eat and drink, to adapt ourselves to any climate or atmosphere.

A great deal has been said in these latter days about a substance alleged to exist in the atmosphere, and which is considered beneficial to health. In order to show how little is known about it, and how those who write about it contradict themselves, and expose their ignorance while endeavouring to define its character, we give a few extracts from an article on " Ozone " in the Scientific American, Feb., 1874 :

" Ozone is generated by lightning flashes !" " Ozone is oxygen in a *negatively electric state.*" " Evaporation of saline solutions disengages ozone." "Ozone is found near the Sea." " Ozone is heavier than oxygen."

Another experimenter in New York, discovered that matches dipped in water and hung up, cleared the atmosphere by generating ozone. We would suggest that they would be more efficacious if burnt. We have read of some other philosopher who declared the best place for generating ozone was a swamp ! We thus have it stated that ozone is a vegetable gas, and as often that it is a mineral gas. Our readers, we think, will be more inclined to side with us when we state the whole discovery to be a humbug. If exhalations for purifying the air are wanted, it is not necessary to hunt up ozone. If the air is poisoned with vegetable or animal matter, then any mineral burnt and given off as gas will purify it, and vice versa.

STORMS.

From what we have advanced regarding the composition, and formation of the atmosphere, and the cause of rain in a previous chapter, it would not be difficult now to arrive at the cause of winds, and storms, and it may not be amiss to lay down some propositions, which may tend to assist those devoted to the study of meteorology, in their predictions of coming weather.

Sir John Herschell in "Good Words," 1864, has a long article on the subject of "Weather and Weather prophets," in which he tries to show that the trade winds, and all others, are caused by the sun attracting the vapour from the earth, making a kind of vacuum, towards which the winds rush from all directions.

Professor Rogers in "Good Words," 1863, also says :—"The wind and every material movement of the atmosphere is primarily, *as we all know*, a consequence of the unequal warming by the sun, of the different latitudes and tracts of the globe's varied surface." This we deny, and believe it is one of those empirical assertions, which are not sustained by facts in nature. That the sun has an influence in causing winds, we admit, for the heat caused by it locally in certain situations so increases the growth and life action of plants,—in swamps for instance,—that vast quantities of vegetable gases are released, and by coming in contact with mineral emanations or gases floating in the atmosphere—near hills for instance—a reciprocal action is sure to take place. The gases will thus mingle, form cloud and rain, and *descend*, causing a vacuum, which attracts the atmosphere from all quarters, thus producing wind. A vacuum, therefore, is not caused by the vapour ascending, as Sir John Herschell says, but by the gases coalescing, occupying less space as vapour and water, and then descending to the earth.

In place of Herschell's and Rogers' untenable theories we present the following :—The "magnetic curves" formed by the opposite poles of a magnet with iron filings—as explained in chapter on Magnetism—suggest a theory of wind currents over the surface of the globe, which accounts for the cause of almost all storms. If there is a powerful current existing between the two poles of a magnet, then as the earth is a magnet and has two poles, there must be a similar current forcing its way from both the North, and South Pole, towards the Equator. This current, therefore drives the cold mineral winds from the poles, into contact with the warm vegetable exhalations from central latitudes, and causes—along with the special reciprocation necessary in each case—all the atmospheric phenomena we are acquainted with.

Thus the calms in the "Horse" latitudes, near the Equator, which continue for days, and during which deluges of rain fall continuously, are easily accounted for, because the two opposing Polar currents endeavouring to meet there, are intercepted by the dense vegetable gases accumulated on either side of the Equator, and by reciprocating with each other, they condense as rain. This of course occurs only at certain seasons. As the earth changes its position with regard to the sun, so does the latitude change, which is to be the scene of the warfare, or rather of marriage, between the opposing gases. Thus in winter, the rainy season is far to the south, near the Equator, because the Polar winds have the mastery over us. As the spring advances, we are enveloped by the belt of rain, and when summer comes the waters have passed over us, and are reciprocating far away to the northward, because the vegetable emanations in summer, owing to the heat of the sun, have now gained the mastery. As autumn and winter advances the

Polar current regains its strength, and again drives everything before it.

Faraday, in his works, indicated some figures of magnetic curves, which show iron filings to arrange themselves round three centres in a magnet—one at either end, and the third in the centre. The central curve could have existed only in his imagination, for the centre of a magnet will not attract even the smallest particle of iron. Maury in his "Geography of the Sea" copies these figures, without ever testing the experiment himself, and in trying to deduce a theory of wind currents from them, he of course fails. Thus here we have another of those false beacons anchored in the ocean of knowledge, which has lured to destruction, or made shipwreck of—in a scientific point of view—one of the most eminent men of science America has produced, and that, too, while he was far on the voyage, and at the threshold of discovery. For there are many facts in his book which indicate, that he only wanted to know of the existence of this great magnetic current from the north to the south pole, to make everything clear to him. But an unquestioning faith betrayed him. Faith although essential in religion, is often as the wrecker's light on the rock-bound shore, or the phantom mirage of the desert, to him who trusts in fallible man.

A curious idea seems to be prevalent, and it is strongly supported by Maury, that all storms are cyclonic, or circular in movement—that is, the winds all blow round a centre. Such occurrences are very rare indeed, and then whirlpools, and waterspouts, are the result. In most of the instances mentioned of these storms, they are traceable directly *to* a centre, but not *round* one.

In the "World of Science" there appeared some years ago, an article on a, so-called, circular storm, which passed over the ship *Solent* at St. Thomas. This is accompanied by a diagram, showing the direction of the wind on the ship, at different intervals during the storm. First there was a tremendous gale of wind :—"The barometer falling, at nine a. m., the storm increased, and at *noon* it was blowing a fearful hurricane, wind STEADY at N. N. W. hf. W.; at 12.15, p. m., it fell a *dead calm*, with a considerable sea; * * * at 12.40, p. m., it became almost dark; and of a sudden, a most fearful and terrible rush of wind succeeded from S. S. E. hf. E., (a directly opposite point) and struck the ship on the port broadside, keeling her over on her beam ends, blowing her fore and mizzenmast right out of her. * * * At one, p. m., the barometer began to rise; 2, p. m., rising gradually, the *wind steady at S. S. E. hf. E.*" This diagram shows the storm to be *round* a centre, strictly according to the accepted theory. But how is it possible to make the circular force correspond with the facts as stated, where the wind blew furiously from only two directly opposite points, *into* a dead calm, which lasted some 25 or 30 minutes, and during the calm at *noon*, it was *almost dark?* How in this case can they account for the calm between the gales, or in the centre? Generally in the centre of any whirlwind, or whirlpool, there is the most violent movement or disturbance. Such a storm as we have described, if circular, must of necessity, have resulted in some such tremendous central commotion. There could not possibly be a calm. If, instead of saying the winds blew *round* a centre, we say they blew *into* a centre, then the whole phenomena can be easily accounted for. A deluge of rain poured continually during the storm, accompanied with intense darkness. This rain, and darkness or opacity, were formed from those contending and combining currents of vegetable and mineral air, which blew from exactly opposite directions; the calm showing where the centre of union, formed a dense dark cloud and deluge of rain, by the reciprocal action. By condensing into rain, these gases would necessarily occupy much less space, and as their horizontal force was spent, they would then form a vacuum, and attract all the surrounding gases, until they were neutralized—only then would there be rest and a clear atmosphere.

In Chicago, United States, during January 1873, we happened to witness a great snow storm accompanied by violent wind. Nothing like it had occurred for years before, and on examining the maps issued next day from the weather office, showing the direction of the wind at different stations all over the continent, we observed that for hundreds of miles around the city, the wind was trending towards, or to one side of it. The most distant points had wind without either snow or rain, those nearest had either rain, or snow, according to the temperature of the locality, while at, and around Chicago, there was nothing but snow.

If those who are stationed at meteorological observatories, in inland towns especially, paid due attention to the nature, temperature, velocity, and direction, of the wind, as sent to them in the reports from the surrounding districts, the exact locality of any storm, might easily be designated twenty-four hours beforehand. The temperature and character of the gases must also be considered, for many while on the way to the storm centre, might reciprocate with each other, and a little storm centre on their own account would be the result.

CHAPTER XX.

ANIMAL, VEGETABLE, AND MINERAL FOOD.

Nothing so much to do with our discomforts as Food.—The body a machine.—Professor Lyon Playfair on Food.—Liebig's classes of food.—Flesh formers.—Heat givers, and mineral ingredients.—Knows nothing of the action of the last class.—Contradictions.—Experiments by scientific men always conducted too loosely.—Animal food only concentrated vegetable matter.—English Navvies and Arabs.—Sepoys and Ghoorkas.—How much an Esquimaux eats, according to Sir John Ross.—Canadian Indians and salt.—Criminals in Holland.—The Scotch and Indigestion.—The action of minerals in the body.

ONE of the most important matters connected with health and comfort, is the food we eat. Nothing else has so much to do with the trifling little annoyances to which we are subject, and those irritations of temper, lassitudes, and headaches, which worry and annoy us. The body is like a machine, which is almost wholly sustained, repaired, and kept in thorough working order—or injured and destroyed—by that which we put into it. Any one who desires to prove this can easily do so. Let him starve himself for a short time, and he will soon find his flesh reduced, and his bones showing their form. Or let him eat but one kind of food for a few days, such as plum pudding, and see how miserable he will feel. Or let him drink a bottle of champagne in the evening, and see how incapable of work he is next morning. Much lesser ordeals than these, however, suffice to make one miserable, and out of sorts.

How necessary is it, therefore, that we should strive to know the *nature* of our bodies, and of the food we eat.

Half the ailments that people are troubled with, could easily be avoided, if a little more attention were paid to the kind of food required, the time of eating it, the quality of each kind actually necessary, and the manner in which it is eaten.

Prof. Lyon Playfair has two excellent articles on "The Nature and Composition of Food" in "Good Words," 1865, which we propose to discuss; but while he gives an immense fund of useful and practical information, he yet admits that neither he nor any other chemist, can explain the working of the mineral matter which we are obliged to take in our food, and how it is we cannot digest anything, or even exist without it.

Liebig, he tells us, classses all food into three divisions : (1) Flesh formers. (2), Heat givers. (3), Mineral ingredients. Many philosophers object to this classification, but Playfair insists on supporting and upholding it. We, however, object to it on much the same grounds as his opponents, viz.,—that flesh could neither be formed, nor heat given, without a mixture of at least two of the classes. If moreover, he knows nothing about the working of one of the divisions, his opinion about the other two are not much to be relied on, because like all other chemical actions, the atoms must work in unison. Playfair again contradicts himself from his own "Table of Food," for he says :—"The Esquimaux live on heat givers principally." Yet we find that some of the greatest heat givers are vegetable substances, such as sugar, rice, flour, and oatmeal, but the Esquimaux does not eat them, he prefers, or rather is compelled to eat meat and blubber, while the people of warm climates eat those very substances which are stated to give the most heat. Why then do the Esquimaux live on blubber. We say on account of their living in a cold mineral region, they require the strongest vegetable food they can find, to counteract the influence of the cold atmosphere. This they get in animal food, which is merely concentrated vegetable substance. Thus it is we can eat more meat in winter than in summer, and the people of hot climates are content with the simplest vegetable diet, along with so much mineral, such as salt, as they can relish.

Playfair also tells us the principal flesh former is animal food. But some of the strongest and largest animals are vegetarians; such as the elephant, the rhinoceros, the hippopotamus, the buffalo, the cow, and the horse; while those animals which supply us with our animal food, all live on vegetables.

We think Liebig's division of food altogether erroneous, for no account seems to be taken of the climate, and that is everything in importance. What would form flesh in the Esquimaux, would only breed sickness or disease in an Arab, and vice versa.

The experiments made by scientific men to prove certain formulas, are all done far too loosely. They seldom think anything of the *nature*, position, and condition of the material, or the locality in which they are performing, or of the nature of the object operated on; and thus their generalizations can seldom be depended on as correct. If a man, therefore, were to diet himself by Playfair's Tables, with the expectation of forming flesh, or losing it, he would probably end by being sick; because, for any such purpose, these Tables are absolutely worthless, until the nature and action of the mineral ingredients in food are explained. When these are known, they may become valuable for reference. Before explaining this action let us give a few facts concerning the nature and character of food.

We explained before, that animals are formed from an excess of vegetable matter. Animal and vegetable food, therefore, we class as one, while mineral forms a class by itself.

From chemical analysis, we find that mineral matter is associated with all substances in their formations.

The close connection that exists between animal and vegetable substances, is shown in the following quotation from Playfair's essay:—"In all kinds of vegetable food capable of affording nutrition, there are certain substances which are not only alike, but essentially the same in composition, as the principles of which the animal body consists. The albumen of the white of an egg, is equally found in the cabbage. The fibrin which forms an important part of blood, and the chief part of flesh, abounds in wheaten flour and in the cauliflower. The casein or cheese which is obtained from milk, is present still more abundantly in peas and beans, and from which the Chinese in reality extract and make cheese for sale." "Human fat also exists in palm oil; the fat of train oil—as got from the whale—is also found in the root of valerian; and the fat of mutton and beef, exists in cocoa beans."

Is it any wonder then that minute animals should spontaneously spring from these vegetable productions when in favourable conditions? They all contain very little mineral matter, and this accounts for what microscopists tell us of brown sugar —that the cheapest kinds are often a mass of living creatures.

As meat then is only a stronger kind of, or concentrated, vegetable matter, it follows that as it is the nature of man to eat both, those who feed on animal food are likely to be the strongest race of men, and so we find it. Playfair says, that one English navvy will do as much work as four rice-eating Arabs. The Sepoys, moreover, who gave us so much trouble in India, were the flesh-eating men of the upper provinces, while our best allies were the Ghoorkas, who are also omnivorous.

As previously mentioned, the season and climate have a great deal to do with the variety, quantity, and quality of the food to be taken. In a warm climate, or in summer, when the atmosphere is charged with vegetable gases, a vegetable diet, accompanied by plenty of salt as mineral matter, suffices for our wants. While in a cold climate, or in winter, we require the strongest animal food we can obtain, accompanied by very little mineral matter, and this we get in animal food.

Thus we find the Arab and the African, using great quantities of salt with their vegetables—the children sucking rock salt as ours here do sticks of candy. The Esquimaux again, as Sir John Ross assures us, can consume daily, 20 lbs. of flesh and oil. Sometimes he will wash down this enormous mass with a quart or two of train oil, and finish with, as a dessert, a dozen tallow candles. They eat very little salt, and their first demand in spring is for a supply of it. The Indians in Canada live in a similar manner; the villages near their winter encampments being besieged in spring for the same article.

Concerning the mineral ingredients of our food, Prof. Playfair says:—" Unfortunately the knowledge of this subject is far from being precise; certain mineral bodies such as phosphate of lime, magnesia, soda, potash, common salt, the sulphates of the alkaline bases, and oxide of iron, are absolutely essential to nutrition. It was formerly a punishment in Holland, to feed great criminals on food free from salt; and they are stated to have been subject to the most loathsome diseases."

"Undoubtedly it is essential to the processes of digestion and assimilation, although we cannot explain more than a few of its actions." "If all the organic elements of nutrition—the flesh formers, and heat givers, were presented to an animal in abundance, in the absence of these mineral substances, the animal would not only cease to thrive, but all nutrition would be impossible." "Our information on this subject is very meagre, and while we recognise the importance of these mineral ingredients of food, chemists do not at present profess to explain their action."

We dare say our readers, by this time, are able themselves to explain the mystery to Prof. Playfair. We have already said that all dissolving processes are caused by chemical action. Digestion is a similar action, and chiefly a reciprocation between the two classes of matter—mineral and vegetable. As our food then is principally vegetable, no motion will take place in it unless it is accompanied with a quantity of mineral matter, when every individual atom is loosened and the whole dissolves away.

In conclusion, we need not be troubled with indigestion, or lack that "mysterious" ingredient, *gastric juice*, if we take a proper quantity of mineral matter and drink a sufficiency of water with our food. It may account for the Scotch, as a people, being so little troubled with indigestion, that the food, especially of the poorer classes, is mainly composed of oatmeal, salt fish, and cheese—substances which, Professor Playfair shows us, have a large quantity of mineral matter in their composition.

CHAPTER XXI.

COAL.

Found to be of vegetable origin.—Prof. Rogers on Coal.—Statements faulty.—Unacquainted with natural law.—Rogers' theory.—Grew in a swamp.—Soaked with mineral oils.—Baked by the earth's internal fire.—A forest makes half an inch of coal.—A tree said to absorb carbon.—Incorrect.—Sir Henry De la Beche and his calculations.—Fallacies about carbon.—How carbon and hydrogen came into the coal.—Our theory of coal.—Prairies.—Charcoal in the seams.—Nova Scotia mines.—Inundations.— No internal fire.—No Baking.—The whole process one of petrifaction.—Coal inexhaustible.

For a great many years, this useful production of nature was very little understood. It was classed among the ordinary minerals, and was supposed to have been made in a similar manner—"*out of nothing by fire.*" Geologists, however, in pursuing their researches some thirty years ago, found numbers of fossil leaves and stems among the coal; and in the shale (a stratified kind of rock always found in connection with coal) appeared a forest of fossilized vegetation. The conclusion was then arrived at, that coal was of vegetable origin.

After this had been discovered, numberless theories were thrown out regarding the way in which these coal fields were produced; but we have not seen one that gives an altogether satisfactory solution of the problem.

Prof. Rogers in "Good Words," 1863, has two interesting articles on "Coal and Petroleum" which contain the latest ideas on the subject; but the statements are so faulty, and show such unacquaintance with natural law, that we propose, firstly, to detail how the popular idea of the formation is incorrect, and secondly, to show how coal is really formed.

Prof. Rogers' theory is, that the coal-producing vegetation grew in a bog or swamp, that it was of colossal dimensions, and as each season's growth decayed, it sunk into a sort of peat. A succession of earthquakes then occurred, gradually sinking the coal measures to their present level. The vegetation being all compressed hard, *petroleum and other mineral oils and juices* soaked through the substance, and the whole field was then baked into coal by the earth's internal fire; those masses which were nearest the interior being baked the hardest, and thus it is we have the hard anthracite, and the common soft coal.

This theory seems very plausible, but as we will show, it cannot be sustained by facts. A forest, it is said, will only make half an inch of coal. How many hundreds of forests then must have grown, or how luxurious must have been the growth, in order to produce a coal field, which, with its shale, is frequently found three or four hundred feet thick.

Prof. Rogers, in order to account for the luxurious vegetation, enters into a speculation, which shows how little is really known of natural phenomena, seen by us every day. A tree, he, as well as all other teachers say, is mainly composed of carbon. This substance it *absorbs* from the atmosphere, and according to the amount of carbon, so is the quantity of vegetation. In the coal-producing ages, therefore, there must have been an incalculable amount of carbon in the atmosphere, for the immense growth of vegetation that then existed could not have sprung up without it. As carbon is deadly to animals, none could have lived at that time. In the "Vestiges of Creation," we find a similar statement to the effect, that Sir Henry De La Beche has calculated, that "if the quantity of carbonic acid gas which is locked up in limestone, and coal, were disengaged in a gaseous form, the constitution of the atmosphere would undergo a change, of which the first effect would be, the extinction of life in all animals." If this is correct, how is it that the work of coal mining is so healthy? Again, the quantity of carbon that is disengaged every day, in a large city where coal is consumed, ought to be so enormous that much injury should result from it; yet we hear of no evil effects. As the coal is being gradually consumed, the whole of the carbon which was once in the atmosphere, must eventually be restored to it again, the earth every year, therefore, ought to be becoming more and more unhealthy, till no animal will be able to live : yet we hear of no appearances to indicate such a change. Lastly, carbon is said to be so injurious to man, that if he breathes much of it he dies; yet we are told that he lives on it, for all vegetables they say are mainly composed of it !

That there is such a gas as carbon, may be admitted, but that vegetation feeds on it, and is mainly sustained by it from the atmosphere, is an impossibility. The only necessaries demanded are heat, good or suitable soil, and water; whatever carbon is found in the tree or plant, has been absorbed from the soil and water, and not from the atmosphere.

The hydrogen in the coal we can account for also. Prof. Rogers says :—" A passing allusion has been made to the absence of any mineral source for the material of the coal beds,

This is a fact patent to every mineralist; and there is another fact disclosed by chemistry, that both the carbon, and the other main ingredient of coal, the hydrogen, could have come together from no sources but the earth's atmosphere and water, and only by the process of vegetable growth or plant life." This was written before hydrogen was generally known, or admitted to be mineral. One source of the mineral, therefore, was from water, but this could not have supplied the coal with the quantity found in it. Prof. Rogers himself gave the cause when he said "mineral oils and juices soaked through the decayed vegetation."

Our view of coal formation is as follows, and we think it will be found to tally with all the facts that have been ascertained relating to the subject, and with true natural law.

In the Western Prairies of America we have a vegetation continually growing, decaying, and growing again. During the time this has been progressing, an immense depth of vegetable soil has been deposited. It is probable that the material of many coal beds, though not of all, was first formed in this manner. Parts of trees are often found fossilized in the coal, but they are all in an upright position, or rather at right angles to the strata of coal. This shows that, comparatively speaking, the growth was of inferior vegetation, such as ferns, grasses, and bushes. They were probably more luxuriant, however, then than now; and in order to account for the great depth of the deposit, many fields by convulsions may be supposed to have doubled on themselves.

In the strata of the Nova Scotian shale and coal fields, are found seams containing charcoal; showing where fires had run over the prairies, just as they do at the present time. As the earth also was in more violent action than now, occasional inundations took place, and sands, sediments, etc., were deposited, in the positions in which we now find them, between and over the coal beds. Taking into account the vast thickness of many coal beds, and the vast quantity of vegetation that would be required to form them, we are inclined to believe it possible that some may have been formed from deposits of vegetable soil, or matter, independent of vegetable growth.

The process which the material undergoes to form coal we will now state briefly.

The first fallacy of Prof. Rogers is that there is any internal fire at all, and this we prove in another chapter. The next is that vegetation, even although soaked by mineral oil, could by any manner of baking whatever be converted into coal. We assert that the whole process was merely one of *petrifaction* As the vegetable deposits were gradually covered over, they would be saturated with the mineral emanations, solutions, and gases, which are continually being formed in the interior of the earth, then, by the chemical action between the two substances, the vegetable matter would thus be thoroughly combined with the others. After a time this would merely harden or petrify into coal. Neither fire nor a baking operation, or pressure, were therefore required. It may be asked, how is it, when the prairies are level, that seams of coal are always found lying at an angle? Because by the mineral gases and solutions petrifying the vegetable deposits, the escape into the atmosphere of other accumulating gases from below is stopped, and in consequence, a force is generated which causes an earthquake. This throws the whole field into the position in which we usually find it. These positions allow a free escape of the ever accumulating gases through the strata; examples of which may be seen at many mines, arising from the fissures in the strata; and sometimes when the *crop* is covered with water, the gas boils through it, and may be collected and burned in the atmosphere.

In conclusion, great fears have been entertained that the coal supply of the world will run out, but we believe, that like every other production of nature it is inexhaustible. By the time the present coal seams are worked out, fresh seams equally as good may generally be found in the same vicinity. Fresh deposits also are continually being discovered, and there is no doubt but that fresh fields are at the same time continually being formed.

CHAPTER XXII.

HOW CORAL GROWS.

Strange Chapter.—Coral insects unworthy of notice.—Misplaced eulogy.—Theories of Coral growth.—The insect monument and tomb.—Not found below thirty fathoms.—Coral found a mile and a half deep.—Coral on the Isthmus of Panama, not made by insects.—The Coral insect a parasite merely.—The cochineal.—How Coral grows.—Millions feeding from one mouth.—Coral grows by budding.—Agassiz on Florida reefs, and argument against Darwin.—Darwin's curious theories on Coral reefs.—Sir John Herschell.—How Coral commences to grow.—The true theory of reefs.—How a gap in a reef was filled.—Coral merely the home of the insect.

The title of this chapter may seem strange, as we have always been taught that Coral did not grow, but was designed and built by small insects. Many are the lessons that have been drawn from their supposed industry, the sermons that have been preached on them, and the lectures in which eminent men have waxed eloquent upon them, but it is our painful duty to inform naturalists generally, that their eulogy is misplaced, that coral insects are no more to be compared to bees, than sand is to sugar, and that they are as unworthy of notice as a common grub or fly.

Coral is only found in equatorial latitudes. In the Pacific Ocean there are islands said to be entirely composed of it, and these appear to have been formed from the bottom to the surface.

On examining the substance, it was observed to be covered with small insects, and scientific men, without much thought, or close research, at once assumed that they built the coral.

Commencing at the bottom of the ocean, we were told, these industrious insects struggled higher and higher from their vast lurid sea green depths, towards the light, until they reached the surface, when, with a devotion worthy of a higher phase of being, they sealed the work with their own bodies, thus making it their tomb, as well as their monument. Later discoveries have proved this to be contrary to fact. *Chambers Encyclopedia* tells us it has been ascertained, that none of the *polypes*, or coral insects, live at depths of more than twenty or thirty fathoms; and that most of them are inhabitants of much shallower water. As coral is found at much lower depths than this, the question arises—what formed it at the bottom of the ocean?

During the cruise of *H. M. S. Challenger*, a piece of coral was brought up from a depth of a mile and a half; and in the newspapers it was stated to be, *unfortunately dead*. If by this it was meant that no coral insects were found in it, then we have evidence that our view of its formation is correct,—viz; by *semi-mineral growth*.

Every one has probably seen coral. It is generally in the form of a tree, with branches; or it is of a circular form, like a sponge, with the appearance of having all sprung from a centre. On the Isthmus of Panama we saw a large solid piece that looked like the trunk of a tree, and on breaking it, we found the formation to radiate from an innumerable series of centres, the whole filled up with an interlacing of stars. The construction was such that it could not have been built by insects, for the openings were so small that they would scarcely allow the point of a needle to enter, far less a polyp to live and grow there; and it afforded in our view, the most beautiful exhibition of mineral growth that could possibly be seen. The forms of the centres resembled the figures of snow flakes,—which is another form of mineral growth, or crystallization,—only they were on a much smaller scale. On being broken, the coral split in vertical layers, and showed the star cavities to continue in long unbroken lines from the top to the bottom of the piece. The only sign of life about it was a number of worm holes circling through it made by the *borer*.

Coral is only a form of mineral growth, and it as surely grows in equatorial waters by natural law, as a tree grows on the surface of the ground. The coral insect is merely a parasite of the coral, just as the cochineal is a parasite of the cactus; and it would be as correct to say, that the one formed the plant, as that the other formed the coral.

We have already explained how the lead tree is grown by suspending a piece of zinc in a solution of sugar of lead,—the formation and growth of the coral is by a similar process.

Coral is composed of material having all the properties of vegetable matter. In the equatorial latitudes, vegetable material is most abundant; the waters, therefore, must also be largely permeated with similar matter. This agrees with the popular theory, for it is acknowledged that the material for producing the coral is in the water, and that it is taken from thence by the insect, and laid with mathematical precision and artistic taste on the fabric. But if we can show how the material may be drawn together without any animal aid whatever, we deal a death blow to another of those sensationalisms of science, which are so destructive to the acquirement of a correct knowledge of the power and functions of natural law.

Various particles of animal and vegetable matter, in the nature of coral elements, settle on a rock or sea bottom. As their numbers increase, they acquire a greater magnetic influence, and attract other particles. But, as atoms, they have the repelling power of the magnet, and like to either the philosopher's tree, or the natural one, they throw out roots on the rock, or sea bottom, and branches in the ocean; looking almost exactly like a leafless tree in winter. The shape, size, and colour, of the formation, is naturally guided by the condition, quantity and quality of the surrounding materials. All coral does not contain insects, and while some, from its coarse nature, may provide convenient abodes for a species of animate jelly, yet, even supposing that this jelly is as high in the intellectual scale as the oyster—which it is admitted not to be—it has no more the power of design, the gift of aspiration, the longing for the light and the glories of the sunshine, than the barnacle, or a piece of seaweed; and no more influence in determining the size, shape, colour and extent of its coral home, than a mouse has over the castle it dwells in.

Compare with this view what the late lamented Agassiz said about *polypes* in his *Cambridge Lectures*, in 1873:—
—" Here then are animals remaining united by the lower parts of their body, but having distinct heads, and distinct internal cavities, yet in which the juices elaborated by digestion in one individual feed the common stock. This may go on till hundreds of thousands, nay millions, of individuals share the result of functions of life and digestion, performed only by a

certain limited number of the community." Fancy millions of stomachs fed from one mouth.

Agassiz again asserts that some species of coral grows by budding. This ought to have been a capital argument against insect formation, but he fails to see it, and moreover states:— "Botanists never look upon a tree as a simple individual, but an aggregate of individuals growing upon the same foundation, and remaining attached to the parent stock." If assertions like these are good for anything, there is no saying where they might end, for a man may, with as much reason, be said to be composed of an aggregate of individuals also.

Agassiz founds a curious argument with coral insects, against Darwin's development of species, which is worth noting. In tracing the formation and growth of coral reefs in Florida, he has shown that eight thousand years are required to raise one of these reefs, or walls, from its foundation to the surface of the ocean, and as there are four wall reefs round the southern extremity of Florida, the first of these must be thirty thousand years old; "and yet all of them are built by the same identical species." "These facts then," says he, "furnish as direct evidence as we can obtain in any branch of physical inquiry, that some at least of the species of animals now existing, have been living over thirty thousand years, and have not undergone the slightest change during the whole of that period." But as we have shown the insects cannot make these reefs, or anything else, Darwin is safe enough yet, as far as that argument is concerned.

Darwin himself, has some curious theories regarding coral insects, and their work, from which he draws conclusions more illogical than any Agassiz has given. Many of the coral islands are surrounded by a reef, a short distance from the shore, and deep water exists between them. Darwin says these islands were originally connected with the reef outside, *but as the islands are continually sinking*, they carried down with them the coral fringing it. While therefore the fringe was carried down, the reef was always brought up to the surface, by the aforesaid industrious insects. Why the insects only worked on the reef, and not on the fringe, is unexplained by Dr. Darwin. Moreover, the depths of the vacant spaces are so great, that if it were true the *islands* were sinking, and had continued to sink along with the fringe, they must have gone out of sight long ago.

In connection with this, Sir John Herschell says in his Essay on *Volcanoes*:—"Dr. Darwin has shown by the most curious and convincing proofs, that the Coral Islands are sinking, and have been sinking for ages, and are only kept above water—by what think you? By the labours of the coral insect which always builds up to the surface." If we only examine the statement for a moment, we will find the arguments are certainly curious, but scarcely convincing.

Suppose the island to sink twelve inches, the insects cannot all assemble below the island, and raise it up again the space of that foot. Neither can they invade the island and form coral all over its surface, a foot high. The only thing they could do, would be to work all around the *shore*, and bring *it* up a foot high. If this had been going on for ages, an immense high wall would be formed all around the shore, and on looking over it we would see a beautiful island, a hundred feet or so below us, looking like a fairy world with luxuriant groves of cocoanut and palm trees. Such romantic scenes, however, are not for this world, and the plain dull truth is simply this, that the Islands are not sinking, that the reefs do not require to be brought up to the surface, and that, not only are the insects not required, but they are also as incapable of the herculean task which the Doctor assigns to them, as a starfish or a sponge.

We assert that the reefs never were connected with the islands, but that they are formed in a similar manner to the bars at the mouths of rivers or harbours, or the sand bars of islands on the North American coast—such as those of Sable Island—which are formed where the wash, and the undertow from the shore, meet. To explain more fully. The bottom of the sea is composed of material and formations similar to the surface of the dry land. Both have their valleys, rocks, hills, and mountains. The higher mountains which rise above the level of the sea must form islands; such as St. Helena, St. Thomas, the Azores, the Bermudas, and Sable Island. What are called coral islands then, are no more formed wholly of coral, than any of the others. The parts formed of coral are probably only the reef on, and outside of them. Comparative quiet is needed for mineral formations, therefore the fringes could not form on shore, as the surf by continually dashing on it, caused too much disturbance. But where the wash and undertow met, there would be an accumulation of the necessary material, and all the tranquility that is requisite. Thus we account for the reef outside the island, and once that reef is properly filled in solid, sufficient to intercept the inrolling sea, another reef by the same action of wash and undertow will be formed beyond, or outside of it; and so on indefinitely.

Chambers Encyclopedia gives an instance of this outer reef having been broken through to allow a vessel to pass out, and says the gap was filled up again *in fifteen years*, by these industrious insects. This incident is introduced to show the rapidity with which they work. But if we are to attribute this work to them, we must also believe them endowed with considerable administrative and governing powers. Orders must have been issued to all parts of the reef, for the assembling of an army to repair the damages; and work in other places must have been delayed in order to finish this great undertaking. We should rather think that they would have been annoyed with the destruction, and left the gap alone. If we could find any such gap abandoned, or not filled up again, then we would admit *our* theory wrong, for the law of nature in repairing the breach could not be stopped.

In conclusion, every production of nature capable of sustaining life has a parasite peculiar to it, and the coral insect is merely one of these parasites making its home in or about the coral, and feeding on the congenial water around it.

CHAPTER XXIII.

VOLCANOES AND EARTHQUAKES.

Another popular fallacy.—The earth's internal fire.—Dr. Mayer's theory. Dr. Tyndall opposed to it.—Dr. Mayer's dogged assertion.—Selfishness of men of science.—Herschell on Volcanoes.—Nineveh and Carthage.—The earth and an egg.—Objection to Herschell's theory.—Explanation of Volcanoes.—Why Volcanoes become extinct.—Coal gas.—Mount St. Helena and Sulphur Springs.—Prof. Mallet on Water and Volcanoes.—Cause of Earthquakes.—Prevention of Earthquakes.—Oil boring in Pennsylvania.—Herschell's extraordinary theory of Earthquakes.—What he knew of chemical action in the interior.—The necessity for scientific men not taking anything for granted.

In this chapter we deal with a popular fallacy advocated by scientific men for many years past; but which has nothing to support it, save their own assertions. This theory is, that the interior of the earth is a mass of molten fire, where everything is reduced to a state of "igneous fluid;" and the crust we live on is only about twenty or thirty miles thick—so that taking the size of the earth into account—the solid substance we are supported on, is not so thick, by comparison, as the shell of an egg, to the matter within it. Then when we fancy that the interior is bubbling and boiling all the time, the wonder is that we have not exploded long ago; or that the Himalayas, the Alps, or the Andes, have not fallen through, and added fuel to the flame.

The theory as expounded by Dr. Mayer in his *Celestial Dynamics*, is as follows :—" Several facts indicate that our earth was once a fiery liquid mass, which has since cooled gradually, down to a comparatively inconsiderable depth from the surface, to its present temperature. The first proof of this is the form of the earth. According to the most careful measurements, the flattening at the poles is exactly such as a liquid mass rotating on its axis with the velocity the earth would possess; from this we may conclude that the earth at the time it received its rotary motion was in a liquid state."

Scientific men will however differ, and we find that Dr. Tyndall does not coincide with this theory. In a note to his book on *Heat* we find the following :—" Prof. Wm. Thomson has recently raised a point which deserves the grave consideration of theoretic geologists. Supposing the constituents of the earth's crust to contract on solidifying, as the experiments thus far made indicate, a breaking in, and a sinking of the crust would assuredly follow its formation. Under these circumstances, it is extremely difficult to conceive that a solid shell should be formed—as is generally assumed—round a liquid nucleus." Dr. Mayer's theory is however strongly supported by Sir John Herschell, Dr. Lardner, and others.

Dr. Mayer proceeding with his theory, says :—" The cooling of the earth must have shortened the length of the day "—as the earth would contract as it solidified, and the contraction would make it revolve quicker on its axis. But Laplace has shown that in twenty-five centuries—the time in which our earth revolves on its axis—it has not altered one five hundreth part of a sexagesimal second. This puzzles Mayer, but he very doggedly asserts, that if they give up the theory of an internal fire—they then deprive themselves of any tenable explanation of volcanic activity. This result however remains to be seen. The assertion of Mayer is of a piece with most of the scientific dogmas of later days, for when facts will not coincide with theories, the ambitious philosophers yet cling to their statements, and by perpetually driving them into people's ears, endeavour to stifle the attempts of others to arrive at the truth. It is to be deplored, that there is displayed among scientific men more of a selfish desire to have their names mentioned in connection with some secondary, and comparatively unimportant pet theory, than is consistent with a true ambition to promote the interests of science.

Sir John Herschell says, in an article on *Volcanoes and Earthquakes, Good Words,* 1863, that they are:—" Unavoidable (I had almost said necessary) incidents in a vast system of action, to which we owe the very ground we stand on ; without which neither man, beast, nor bird, would have a place for their existence, and the world would be the habitation of nothing but fishes." The reason he assigns for this is, that the land being every day washed by the tides and rivers, is having vast quantities of its material gathered and deposited in the ocean ; and this is so constant and universal, that if there were no counter action by earthquakes to raise the land, the earth would be one vast ocean ! That earthquakes raise vast tracts of country occasionally, we admit, but that they have the influence which Herschell attributes to them, we deny.

Volcanoes and earthquakes according to the popular theory, are thus caused by the central heat of the earth. Herschell attempts to prove this supposition on different grounds, from Mayer, by saying that when we descend into the earth, the atmosphere becomes gradually hotter, till—supposing the rate of increase to continue the same—at twenty or thirty miles the crust of the earth would be found red hot. Thus the English commissioners have fixed four thousand feet as the limit for the depth of the coal mines. The health of the colliers, it is supposed, would suffer, should the hope of gain induce mine owners to sink their shafts below that depth.

The objections to this theory are very numerous, although the number of hot springs, and volcanoes and earthquakes, would seem to support it. The main objection is, that Herschell and others have calculated the heat of the interior, from the surface inwards; they have never calculated what the effect would be on the surface of the earth, from the amount of heat they say is inside of it. If an egg shell and the meat within it, is a correct illustration of the relative thickness of the crust of the earth to the molten matter within, then we wonder how hot the egg shell would be, if the contents were only boiling hot. Any calculation at all would show us, that if our crust was only twenty miles thick, the heat would be so intense on the surface, that, far from the world being only inhabited by fishes, our oceans would be dry, our river beds empty, and neither man, beast, fish, or plant, could have any existence whatever. The whole theory is such an untenable one, and so much out of harmony with natural law, that it only remains for us to give a feasible explanation of the cause of volcanoes and earthquakes, to have it rejected altogether.

Firstly, then, with regard to Volcanoes.

They are mountains, from the tops of which issue—when in

activity—smoke, flame, ashes, and lava; and hence it has been argued that if fire issues from the crater, there must be fire inside. This, however, need not of necessity follow. The whole phenomena connected with both volcanoes and earthquakes, are caused by chemical action. The intense flame witnessed at the mouth of a volcano, is caused by the ignition of the vast volumes of gases which are issuing from it. The flame does not extend inside the mountain, for it is subject to the same conditions as our coal gas is. When we light the gas in our rooms we have flame, but it does not extend inside the pipe or burner; in fact there is a space outside, between the burner and the flame, that will not ignite. This is owing to the fact, that until a sufficient quantity of oxygen is combined with the hydrogen, there can be no combustion or flame. If we could introduce coal gas into a vessel containing other mineral gases, we could not light it. In a similar manner, there can be no flame in the interior of the mountain, for there is not sufficient oxygen inside to induce a combustion.

Under the surface of the mountain, in the interior of the earth, there are immense stores of mineral materials, such as sulphur, nitre, salt, iron, etc., saturated with water, which chemically act upon each other. By this action, they are continually dissolving, reforming and generating gases, till they accumulate in such quantities that they cannot be confined; they then burst with terrific force, combine with the oxygen of the atmosphere, and igniting, burn until the explosive and combustible gases are exhausted. The dissolving and reforming action with water may then continue, and fresh outbursts may occur until the whole material is changed, then the volcano becomes extinct. Thus we find numberless extinct volcanoes all over the globe, a result which would never ensue, were they—as we are told—funnels or chimneys for the molten fires within. A good illustration of this exists in California, where Mount St. Helena, before mentioned, being once an active volcano, has long been silent, but the evidences of its violent action are to be seen in the wonderful Petrified Forest, while the remains of its once extensive stores of gas producing matter, now burn away in insignificant puffs in the Great Geysers, and the Hot Sulphur Springs; the whole phenomena suggesting the picture of the charred and dying embers of an extensive conflagration.

The lava and burnt earth found at the "crop" of the coal at the Albion Mines in Nova Scotia, show the site, where the gas generated from the action of water in the mine, at one time formed a volcano.

To show that we are not altogether alone in our statements regarding the cause of volcanoes, we may state that Prof. Mallet of England also maintains, that there are no volcanoes without the action of water.

EARTHQUAKES.

Earthquakes are caused similarly to volcanoes. The interior of the earth, as we have said, is saturated with water, and forms a grand magnetic battery, continuously in action, dissolving and reforming its mineral matter—which action produces the mineral gas we have so frequently referred to—and throws it off through the strata of rock, principally from the poles of the earth, and mountain tops as poles. But where this gas is confined, it accumulates, and finally generates a force sufficient to produce an earthquake, and to rend portions of the earth asunder.

If Artesian wells were bored to a sufficient depth in those places that are subject to earthquakes, we are inclined to believe that there might result a comparative immunity from them, for those gases which are dangerous might thus be provided with a vent to pass off without accumulating.

Volcanoes which are always smoking seldom break out with great fury. The danger lies in the apertures closing up for long periods, and thus preventing the escape of accumulating gases.

Some suggestive illustrations of the force of these gases in the earth, occurred while boring for oil in Pennsylvania. In many instances, after boring for a few hundred feet, a cavity would be tapped containing gas, oil and water, and the pressure on it had been so great, that the boring machinery and everything connected with it, were blown to a great height above the surface. Thus compressed gases were relieved, which in time might have generated an earthquake.

Sir John Herschell has an extraordinary theory regarding earthquakes. It is to the effect, that the continual washing away of the land into the sea causes some parts of the earth to become top heavy, and then they fall, making a crack through which the molten fluid escapes.

Notwithstanding all these strange doctrines, Herschell was quite conversant with the chemical action going on in the interior, and with the force of accumulated gases, yet he evidently refused to accept the natural deductions from these truths, and would rather propound a sensational explanation than a simple one. He says:—" There is no doubt that among the minerals of the subterranean world, there is water in abundance, and sulphur, and other vaporizable substances, all kept subdued, and repressed, by the enormous pressure." "But let the pressure be relieved, and they make their way to the surface, expanding as they rise, till they burst in great power."

This admission shows how necessary it is for every man of science to beware of taking any doctrine for granted, unless they have sifted it in every particular, and made it stand the test of their own observations, experiments, and calculations; for it is evident that had Herschell allowed his own common sense to guide him, he would probably have discovered, and caused to be accepted long ago, the true explanation of volcanoes and earthquakes.

CHAPTER XXIV.

THE TIDES.

The regularity of the Tides.—The influence of the new and full moon on the Tides.—There must be one grand cause of the Tides.—This is pressure, not attraction.—Cause of variation in the Tides by the position of the moon.—Formation of the Land.—Winds.—Lardner's theory of the Tides.—Its fallacy shewn.—The earth ought to be approaching the moon.—Facts to be remembered.—The Plane of the Ecliptic.—The effect of pressure on the atmosphere.—The Tides caused by pressure in passing the Plane of the Ecliptic.—The moon's atmosphere.—The Tide in the Mediterranean.—The Bay of Fundy Tides seventy feet high.—Ram Pasture.—Rise of two feet in three miles.—The repelling forces control the Tides.

THE cause of the tides is an interesting subject of enquiry, and yet there are few who attempt to explain it. The periods of the tides are so regular, that they may be determined for months, or even years, before their recurrence; yet again they are never so regular that it can be said the waters will rise to such a point, and no higher. The rise and fall of the tides are the same at scarcely any two places. In the Mediterranean the rise is only about one foot, whereas in the Bay of Fundy, it is sixty or seventy feet.

When the moon is at the new and the full, the tides are unusually high. When a storm also is blowing on the coast from the ocean, the tide in the harbours exposed to it, is elevated several feet higher than usual.

We glean from the above facts, that while there must be one grand regular cause of tides, there are several peculiar influences to which they are subject, and which cause them to vary in different localities, and at different seasons. This grand cause we believe to be induced by *pressure*. The variations to which they are liable are produced :—Firstly, by the position of the new and full moon—secondly, by the formation of the sea coast—thirdly, by pressure of winds.

Before explaining our views, let us review the old established theory and show that the grounds for accepting it are not conclusive.

In consulting various authorities, we find that *attraction* is given as the cause of the tides, and that it is the sun and moon which are the reputed possessors of the influence.

In Dr. Lardner's *Science and Art* we find an explanation to the following effect, of the two tides a day, one on either side of the earth at the same time :—Let us, for instance, imagine a diagram of the earth, with the four cardinal points marked on it. If the moon is directly above the north point, Lardner says, the waters will be heaped up there by the moon's attraction, while on the east and west sides there will be low tides. But as the moon could not attract the waters on the south end also, the high tide there is caused by the earth being *drawn in* from the water.

A more extraordinary theory has seldom gained ground, and we wonder how it was ever accepted, for it will not bear even the slightest examination. In the first place, how the moon being the smaller body; could attract the earth the larger body, is a mystery ; for according to all rules of attraction or gravitation the greater always influences the less. Secondly, the same rule applies to the question, why does the sun, the larger body, not exercise a greater attractive influence over the tides, than the moon? Thirdly and lastly, If the moon draws the earth in, or the earth "recedes" from the water at the side more distant from the moon, it follows, that as the tides are rising and falling continuously from one part of the globe to another, the earth is gradually receding from the water hour by hour, and accordingly ought to be approaching nearer and nearer to the moon every day. This however is not the case, and would be indignantly denied even by Lardner himself; but it is the only logical conclusion to which his theory leads.

Having thus shown that the tides are not caused by attraction, we now explain how, according to our views of the law of nature, they are caused by *pressure*.

In order to understand the assertions clearly, the following facts should be well kept in remembrance, viz.: That the moon has an influence on the tides. That the highest tides occur when the sun and moon are on the same side, or one on either side of the earth. That high tides follow the meridian by two or three hours, and in many places by ten or twelve hours, according to their position on the coast. That the whole rise of the tide is, on an average, only a height of four or five feet. And that while the earth moves from west to east, the tides go from east to west.

Recalling our arrangement of the solar system expounded in previous chapters, we have a number of planets revolving round the central planet, the sun. The planets all revolve on one level, called the Plane of the Ecliptic. Each planet has an atmosphere that extends till it meets the atmosphere of the planet next it. We know also this fact in connection with our atmosphere, that if it is pressed by currents of wind, it in turn presses whatever is opposed to it. On water, therefore, we find that wind storms will sometimes raise the tide two or three feet higher than usual. If our atmosphere, consequently, is pressed from without,—that is, by any other atmosphere—the oceans are impelled to show evidence of such pressure by their rise and fall.

It will be observed, then, that as the earth revolves on its "axis," its atmosphere must be pressed by the atmospheres of the other planets whenever it crosses the Plane of the Ecliptic, as they are more confined at that point than any other. As the earth revolves once on its axis daily, the whole of the globe's surface must be exposed twice to the pressure of the other planets' atmospheres on the Plane of the Ecliptic. We have two tides daily. What is the natural inference to be drawn from these facts? *Nothing less than that the pressure of other atmospheres on the atmosphere of the earth, as it crosses the Plane of the Ecliptic, causes the tides.*

But it will be argued that we make nothing of the influence of the moon, which undoubtedly has some action on the tides. We admit the fact of that influence but its effect is apparent only when the moon is at the new or the full. The tides are then *much higher than usual*, and this arises, in a similar way, by the moon crossing the Plane of the Ecliptic, and adding the pressure of its own body and atmosphere to the influence which is felt there already. Although it is denied that the moon has an atmosphere, such a condition would not be in accordance with the existence of any other known body, and according to atomagnetic or natural law it could not exist without one.

It may be asked, how do we account for the tide following the meridian? The *attraction theory* explains it, on the ground that *inertia* keeps the mass of water from immediately rising, and obeying the attraction of the moon. We may also answer that *inertia* prevents it from yielding at once to the pressure of the atmosphere; and that just as the ocean billows are largest when the tempest has nearly blown itself out, so the tide wave is highest after the pressure is over.

In answer to the query—What makes the tide vary so much in different localities? We answer, that this is all due to the formation of the land at the points of contact with the sea, or its position with regard to prevailing winds.

The reason that there is scarcely any tide observed in the Mediterranean, is owing to the winds blowing three fourths of the year into the bay at the mouth of the Straits of Gibraltar. This causes a strong current to flow through the Straits into the Mediterranean, and as the waters are thus raised higher than the general sea level of the Atlantic, and have no other outlet, they flow back again to the Atlantic by an undercurrent. This we proved by experiment in the year 1830, on board the brig *Clarence*, John U. Ross, master.

In the Bay of Fundy, which is a long straight bay, with a very wide entrance, tapering to a narrow point like a funnel, there is a higher tide than at any other place in the world. This is caused by the tidal wave striking the shore, and running along the North Atlantic coast into the wide mouthed bay, and, as the funnel is continually narrowing, the waters are pressed far above the sea level, because they cannot fall back owing to the heavy pressure of the currents behind urging them on. Tides of sixty and seventy feet thus occur daily at the head of the Bay. In Cumberland Basin, at the head of the Bay of Fundy, near Sackville, New Brunswick, the tide flows up Tantamar river, and round a peninsula called Ram Pasture. This peninsula is only connected by a narrow neck of land fifty yards wide, and is three miles in circumference from neck to neck. This neck is very flat and level. When, therefore, the tide in rising has encircled the peninsula, the waters flow back over the neck into the river again, *falling into it a distance of two feet.* Thus showing that in three miles the water has risen a height of two feet.

It will be evident, therefore, that great inaccuracies must prevail from similar causes in certain parts of the world, regarding the height of rising ground above the general sea level.

In conclusion, while we have great faith in the mutually attractive forces of nature, we have no reason to believe that they cause the tides, but rather that the influence exerted over them arises from their counterpart, the repelling forces of nature, which have here sole control.

CHAPTER XXV.

THE GULF STREAM AND DEEP SEA CURRENTS.

The cause of the Gulf Stream.—Dr. Carpenter's theory.—Oceanic Circulation.—Experiment with glass trough.—No Comparison.—Strength of Polar Currents.—Channel between Faroe Islands and Shetland.—Dr. Wyville Thomson differs from Dr. Carpenter.—Reciprocal circulation of Water and Air.—Beautiful theory of Atmospheric Circulation overlooked by Dr. Carpenter.—What causes the cold deep waters.

AFTER being one of the wonders of the world for a long time, and a mystery which men of science in vain tried to unravel, we have now come to regard the Gulf stream as a phenomenon very easily explained by the various natural causes which give it birth.

The trade winds, which blow almost constantly from East to West, press and force the waters of the Atlantic into the Gulf of Mexico, and these along with the rivers running into the Gulf, raise the waters higher than the general sea level of the Atlantic. As the narrow Strait of Florida is the only free outlet from the Gulf, a rapid and permanent motion is given to the water, which it maintains, in a greater or less degree, for hundreds of miles; and thus the Gulf Stream is established and kept moving.

Dr. Carpenter while describing the cause of the Gulf stream in a somewhat similar manner, sees in it only a particular instance of what is constantly occurring all over the Atlantic and Pacific Oceans. He has a theory of a general oceanic circulation which he is endeavouring to have accepted as an established doctrine of science, viz.: That there is a continuous undercurrent of cold water throughout the oceans, from the Poles to the Equator, and a surface current of warm water from the Equator back again; thereby equalizing the temperature all over the world, and bringing every part of the waters to the surface, to purify and make it fit for the preservation of animal life in the deepest sea beds. He has made his observations from a series of deep sea soundings, in which he has found that the sea bottom, even at the depth of two and three miles, has averaged thirty degrees by Farenheit's Thermometer. His opinion is, therefore, that the water could not be so cold unless it came from the Polar regions, inasmuch as cold water always falls to the bottom, while warm water floats on the surface.

In order to illustrate his assertions he makes an experiment with a glass trough, six feet long, one foot deep, and one inch wide. This is filled with water. At the surface of one end he fixes a piece of ice, to represent the pole, and at the other he applies a bar of heated metal, for the equator. Dropping some blue colouring matter into the polar end, and some red at the equator, a circulation is at once seen to be established. The cold blue water sinks, then creeps along the bottom till it reaches the heated end, is there warmed, rises to the surface, and returns to the pole again; while the reddened water flows along the surface to the pole, sinks and returns just as it is expected to do. But this is not a fair illustration, because what would occur in six feet of water, would not well apply to a distance of three or four thousand miles—the experiment is not a comparative one. The ice and the heated iron almost touch one another, whereas in the ocean there are thousands of miles of water of a medium temperature. If the trough were a mile long, by the same depth and height, it would afford a better comparison, and, of course, it is easy to conjecture that the two temperatures would never meet to form a current.

This polar current Dr. Carpenter says is so strong, that in many places it has rounded the stones at the bottom of the sea, as in a channel between the Faroe Islands and Shetland. But it only requires one to reflect how much force is necessary to roll stones over one another in order to round them, to be convinced that this explanation of their form is not the correct one. Besides, at another place he says the motion is only a *creeping* one.

Dr. Wyville Thomson, one of Dr. Carpenter's colleagues in Deep Sea Explorations, says in his late book on the *Depths of the Sea*, that he differs from him in his theory of oceanic circulation, and does not think that the facts he has given warrant him in arriving at the conclusions he draws from them.

Moreover, it is not a matter of necessity that there should be such an oceanic circulation. It is enough for the purpose of keeping the water pure, and preventing stagnation, that there is a continuous reciprocal or chemical atomic action between the upper and lower strata of water, throughout the whole body of it. This we believe to be the true producer of the oceanic circulation, for its action is not confined to the ocean, but pervades the whole system of our planet. For instance, water percolates to the interior of the earth, and by its chemical action, dissolves minerals and produces gases which find their way to the surface; there they meet other gases, which, by combining form fog, clouds, rain or snow, that descends as water, and percolates back again. On the surface of the earth vegetation is forced out, as from a centre, by the magnetic force of the escaping mineral gas, and this again exhaling a vegetable gas into the lower atmosphere, reciprocates with the rising mineral gases and forms dew, or cloud and rain.

Or, let us take for consideration the bottom of the ocean. The mineral gases from the interior of the earth are forced into the water, and thence through the whole mass of it; thus the warm vegetable atoms which are at the surface, left from evaporation and from rivers running into it, are reciprocated with, or neutralized and purified. The surface of the water, again, reciprocates by evaporation with the lower atmosphere, and that again with the higher, thus forming, as said before, rain, and cloud etc. Throughout the whole extent of our planet, therefore, from the interior of its solid body to the utmost verge of the atmosphere (for we show that meteors are caused by reciprocation also) there is a continuous action ever taking place, forming, dissolving, and re-forming, and keeping up a universal *reciprocal circulation* which prevents stagnation of any kind.

For any one to suppose that nature works in the slow fashion of causing all the impure water distributed over the surface of the ocean, to creep to the Poles before it can be purified again, is egregious folly; for it is evident on the face of it that such is not the case. To bring the matter home, does Dr. Carpenter really mean to say, that if we pour a pail full of slops into the sea, it will require to float onward to the Arctic Pole before it can be changed? We should certainly say not, for in the slops and in the salt water, are the two different classes of atoms, and by each attracting its like, they neutralize each other; this process, as we have said, is continually proceeding all over the surface, and throughout the whole body of the oceans.

To press the question still further, the atmosphere is also being charged continually with all sorts of gaseous impurities induced by the heat of the sun, and given out by plants and animals. Does Dr. Carpenter mean to affirm that we must wait for a storm to blow them away, or must they also travel to the Poles to be purified. In either case a plague would probably result before we could be relieved by these modes. The phenomenon of dew shows us that the reciprocal circulation is continually in progress, and that even when there are no winds blowing, the purification of the atmosphere is accomplished by the chemical action of the atoms.

This leads us to ask the question why Dr. Carpenter has not endeavoured to found a general system of *atmospheric* circulation, on the same principle as that of the oceanic. If the law is good in the one instance, it ought to hold good in the other. He would then have made his theory complete, for the cold Polar current of air being mineral, would be the highest, and this would form an upper current from the Poles towards the Equator, while the warm vegetable atmosphere being denser, would travel on the surface from the Equator along with the warm current of water to the Poles, the one thus helping along the other. A beautiful and complete theory of ocean and air currents would thus be enunciated, the motions of the one accurately fitting into the motions of the other. But unfortunately, he only managed to grasp the least acceptable theory, for as we have shewn in the chapter on Atmosphere, there is more than a grain of truth in the theory of the upper atmospheric current, while there is none in that of the oceanic.

While we deny that there is a general oceanic circulation, such as Dr. Carpenter describes, we admit that the condition of the water and the position of the land, may, in many instances, be such as to induce currents, that, from a superficial examination, may lead an observer to suspect they form part only of one general movement.

It may be asked, how do we account for the extreme coldness of the deep waters if it is not caused by the Polar currents? We answer by asking and answering another question. What causes, or how do we account for, the coldness of the Polar sea? If the conditions for causing coldness are the same on the sea-bed at the Equator, and in the Arctic ocean at the Poles, there is no necessity for the supposition of a current at all. This we endeavour to show.

The gases expelled from the interior of the earth are mostly mineral in character. As the earth is a magnet, its gases, are given off principally at the Poles. Ice and snow being composed largely of mineral substances, are therefore caused by the mineral emanations from the Poles of the earth combining with suitable vegetable particles of the water. (This mode of ice formation accounts for the granulated, and not stratified character of icebergs.) But these mineral emanations are not confined to the Poles, they are exhaled all over the earth's surface, and wherever we find a place free from vegetation, or the influence of vegetable gases, there we may observe it more particularly. Thus the tops of high mountains are cold and covered with snow, while in the depths of the sea where no vegetation exists, and the warmth of the sun does not reach, we also have the cold mineral exhalations acting upon the salt water, and producing the so-called Polar current.

CHAPTER XXVI.

COMETS.

Very little known about Comets.—Facts about them.—Jupiter's influence on them.—Comet of 1680.—Herschell's description of it.—The movements of a Comet different from a Planet.—All the heavenly bodies, Magnets.—The motions of Comets explained on this theory.—How Comets are made periodic.—Encke's and Biela's Comets.—The atmospheres of Comets.—Their tails.—Their purpose.—Are they inhabited?

In writing this chapter on Comets, we do not intend to enter into a full history of them, for that can be obtained in any book on Natural Science; neither do we pretend to give a correct account of what they actually are; but as they come within the influence of atomagnetic law by entering our system, we propose to offer some explanation of the phenomena and movements which they exhibit, and also a suggestion as to their character.

On an average, it is said, two or three comets visit us every year; but most of them are too small, or too distant to be seen by the naked eye, or are situated in such a position with regard to the earth that they cannot be observed.

The following are a few of the particulars connected with comets, which may help us to an explanation of their nature:—

First.—They all seem to be attracted towards the sun; but none have ever been absorbed by it.

Second.—Jupiter sometimes attracts a comet on its way towards the sun, and delays its movements, but always repels it again. None of the other planets have been reported to influence comets in any way.

Third.—Some comets, such as Halley's, Encke's and others, return within a certain number of years; but by far the greater number are never again noticed or reported.

Fourth.—While a large number of comets have luminous tails, most of them have none.

Fifth.—Comets do not keep on one plane, nor move similarly to the planets, but they appear to come from any, and every direction.

Sixth.—Sometimes a comet when it approaches the sun is shot straight back again with tremendous force. This was the case with one in 1680, which Sir Isaac Newton saw, and which he endeavoured to explain by the law of Gravitation; but he could not account for the force which repelled it from the sun.

The comet of 1680, Sir John Herschell says, was perhaps the most magnificent one ever seen. It appeared from November, 1680, to March 1681. It was not very bright at first, but as it approached the sun it grew more brilliant. "When within one-sixth of its surface, it *turned violently round*," and was shot back in exactly the opposite direction with such tremendous force, that in four days it was as far distant from the sun, as in approaching it, the comet took twenty-eight days to travel the same distance.

Seventh.—It has not yet been ascertained whether the nucleus of a comet is solid or not, but from its intense brightness, it is supposed to be so.

Thus in the motions of comets we have an entirely different movement from that obeyed by planets. A planet circles round the sun in a regular orbit, while a comet goes apparently beyond the bounds of the solar system; and when it does come within it, it speeds towards the sun at a tremendous rate, and after approaching very close, it is generally repelled again with greater force than ever. This is contrary to the so-called principle of Gravitation, according to the law as promulgated by Newton's disciples in the present day; for they say that if the earth was to approach as near the sun as a comet does, it would assuredly fall into it.

Believing all the heavenly bodies to be magnets (many *savants* are now urging the propriety of the theory)—and Atomagnetic law to be universal—the movements of comets are explained at once.

Supposing a comet, on its journey through the universe, be drawn within the influence of our solar system, it is attracted by and speeds directly to the centre of our magnetic attraction, the sun; but as soon as it comes within the influence of our luminary's denser atmosphere, it undergoes a change. What is the nature of the change? If two different sized magnets are brought together, the greater will instantly reverse the polarity of the lesser. Thus it is with the comet, the larger body being the sun, it instantly reverses the polarity of the former; dissimilar poles are thus brought together, and as similar poles repel, the comet is driven from the system with irresistible force. It would then never return, unless it came under the influence of another sun, which reversed its polarity, and sent it back again.

Those small comets such as Encke's, and Biela's, have been rendered periodic by being hemmed in by the attraction and repulsion of the planets and the sun, and made to revolve on the Plane of the Ecliptic.

That comets have atmospheres may be accepted as correct, for their bright light could not be caused without it. The fact of their having an atmosphere, also, should reassure us that they can never do any harm by coming in contact with us; for besides the magnetic repelling power each body possesses, our atmosphere is elastic enough to preserve us from close contact. It might be worth while, however, to inquire whether the pressure of any comet which approached very close to us, especially on the Plane of the Ecliptic, had any influence on the tides.

The luminous tails of the comets may be attributed, we believe, to a cause which we have already suggested in explanation of the transparency, or opacity, of atoms. We find that although water is clear and transparent, yet if blown into foam, it becomes white and obscure, simply because the polarity of the atoms is disarranged and in motion. Thus at night, off the coast of Mexico, we have seen the crested billows as they folded over one another, flashing brightly as they curved into foam, spangling the whole surface of the sea with the dazzling spectacle of an endless succession of phosphorescent beacon lights. So in a similar manner, we believe, that these vast luminous envelopes, as they are called, stretching it is said, for millions of miles, and which strike even the most cultured observer with awe, are caused by the atoms of the material filling all space—which are arranged according to their natural atomic position, and are transparent—being disturbed by the different atmosphere of the comet in passing, and have their poles deranged, thereby causing a motion and an opacity, which in the darkness we see as light. This displacement continues till the comet is no longer able to influence them, and then they assume their natural position again. The tails of comets are long, or short, or invisible, according to their position in regard to the earth.

The light of the comets is caused in a similar manner to that of the planets, by a reciprocal action of the magnetic force in each, acting on their atmospheres. What may be the purpose of comets, and the object they accomplish, are questions which, in the present state of our knowledge regarding them, are unanswerable, until we can find out the extent of their journeys, and what other heavenly bodies they may circle round.

Whether they are inhabited or not, is also a question to be decided when there is more discovered about the solid nucleus, but the probability is that they are, like our Planet, inhabited by some description of beings.

CHAPTER XXVII.

METEORS.

Strange theories regarding them.—Sir Wm. Thomson's.—Seed bearing Meteors.—Prof. Newton on November Meteors.—No orbit of Meteors.—Meteors caused by pressure and reciprocation.—Dr. Sorby the microscopist on Meteors.—Prof. Graham on the Leonarto Meteor. —The great November showers caused by a Comet.—The yearly and ordinary Meteors caused by pressure.

For a long time no attention whatever was given to the study of meteors, but lately there has arisen a great interest in their movements; and, as usual, the most extravagant theories are promulgated regarding them. Some physicists say they are fragments of broken worlds travelling through space; others that they are ejected from the sun and moon; and others that they fill all space and rush toward our sun, where they act as fuel to keep the fires burning. But the most extraordinary one of all, is that theory, before referred to, suggested by Sir Wm. Thomson, that *meteors are seed bearing*.

The following are a few of the facts concerning them.

While meteors and shooting stars may be seen nearly every clear evening, there are particular seasons when they are more numerous than usual. On the 9th and 10th of August, or the 14th of November, there are often to be seen magnificent displays. This fact has led some astronomers to suggest, that there must be a ring, or several rings of meteorites which travel round the sun in an orbit, bearing a similarity to the orbit of our earth. But the great objection to this theory is, that they move in a contrary direction to that of the earth and the other planets. Another peculiarity is, that in the history of all the aerolites contained in museums, only one was observed to fall on a " star shower" date. Professor Newton, of Yale College, America, also states, from observations taken with the spectroscope, that the shooting stars of November consist of more inflammable material than those of other meteoric showers. All these facts tend to show the improbability of an orbit of meteors.

Sir Wm. Thomson's theory of the seed bearing meteors is so improbable, and has so little apparent evidence to support it, that it may be dismissed at once. We would, however, call attention to the argument which he brings forward, to show how it is possible for two worlds to have collided, and distributed the seeds. He says:—"It is as sure that collisions must occur between masses moving through space, as it is that *ships steered without intelligence directed to prevent collision* could not cross and recross the Atlantic for thousands of years, with immunity from collision." We are astonished that such a student of science should have studied the mechanism of the universe to so little purpose, as not to have discovered the perfection of every movement connected with the heavenly bodies; and compared with which, human action or skill is as perpetual blundering. That any leader in science at the present day, should consider it possible that planets, and other similar bodies, are not gifted with attractive and repulsive properties and atmospheres, calculated to keep them from colliding; only shows how little is really discovered by scientific men about nature and her laws. Besides he can point to no direct evidence of any collision having ever taken place; and of all the numerous erratic comets which have passed through our system, not one is known to have collided with a planet.

. Our theory of Meteors is as follows :—The earth in its daily journey through space, has its atmosphere pressed by the atmospheres of other planets, (as shown in the chapter on Tides.) The outer or upper surface of the earth's atmosphere, is composed of metallic atoms. If, then, by the pressure or contact of the atmospheres, two of these metallic atoms should unite, they must immediately influence and attract other atoms. They then accumulate, become more and more dense, and fall towards the earth, increasing in bulk, force of attraction, and also in velocity, till they come into the oxygen atmosphere near the surface of the earth; when the reciprocal action between the two classes of atoms takes place, and the friction is so intense that combustion and light is the result: they may either then explode and be dissolved into a gaseous condition again, or fall as solid meteoric stones to the earth.

To support our assertions we have three facts, besides Prof. Newton's statement already referred to, to bring forward in proof.

The first is, that the microscopist, Dr. Sorby, of Sheffield, England, has arrived at the conclusion, from the examination of a large number of specimens, that the structure of meteorites cannot be explained in a satisfactory manner, except by supposing that *their constituents were originally in a state of vapour.*

The second is, that the late Prof. Graham said the "Leonarto" meteor had so much hydrogen in its iron, that the inference was, "*the meteorite was extruded from a dense atmosphere of hydrogen gas.*"

The third is, a discovery which has lately been announced by Schiaparelli, *that the great star showers of November, which occur every thirty-three years, have been hitherto preceded by a comet.* There is no doubt, then, but that it is the great pressure of a comet's atmosphere on the atmosphere of the earth which causes the atoms to combine and to form meteors. We have also no doubt, that the yearly display in August, and November, will be found traceable ere long to a similar cause; or to some combination of the atmospheres by which there is more pressure exerted at one time than at another. The meteors which we see on ordinary occasions, are probably only caused by the constant daily effect of pressure on the atmosphere.

CHAPTER XXVIII.

AURORA BOREALIS.

Visible at both poles.—Mairan on the extent of the Sun's atmosphere.—Lardner on Auroras.—M. Biot on Polar Volcanoes.—Distance of Auroras.—Seen by Aeronauts below them.—Facts.—Caused by mineral emanations from the Polar Latitudes.—How they affect the compasses.—Why seen on Calm evenings.—Dew.—Cause of colours.—Similarity between Auroras and Meteors.

AURORAS are visible in the direction of both the Northern and Southern poles—various explanations have been given of them, but they are all open to objection.

A curious one is given by Mairan, who supposed the aurora to proceed from the intermixture of the *far extending atmosphere of the sun with that of the earth.* While we do not altogether coincide with it, we are glad to get an incidental confirmation of our theory of tides and atmospheres from him.

Lardner says:— "Although the complete explanation of the aurora has not been accomplished, the electricity and magnetism of the earth and its atmosphere must now be regarded as its source." This is partly correct, but as he does not apprehend the properties or mode of action of either, of course he could not understand the theory in its entirety.

M. Biot supposed that metallic clouds thrown out of polar volcanoes were the cause of auroras, and this accounted for their disturbing the magnetic needle. But the disturbance of the needle can be accounted for in a more simple manner.

Some philosophers asserted that the auroras are thousands of miles above the surface of the earth, but the most generally accepted theory is that they are within our atmosphere; for aeronauts have frequently been elevated above them.

The principal facts to be borne in mind concerning auroras are these:—

Firstly, that (in our latitudes) they are always seen to the northward of us, never to the south—even although persons to the south of us may see auroras between us and them.

Secondly, that when they are visible, the evening is usually calm and clear.

Thirdly, that they influence the compass needle and affect the telegraphic lines.

Fourthly, that they have different shades of colour.

We have already shewn by the experiment of the "magnetic curves" that there is a magnetic force or current continually extending from the Poles towards the Equator. The poles of a magnetic battery if introduced into a vase of oxygen will exhibit light, and we have proved that sunlight is produced in Nature from a similar action. We have, moreover, shown, that when the poles of atoms are disturbed, opacity is the result in daylight, and light is the consequence in darkness. We have, therefore, two explanations open to us. The magnetic upper current issuing from the north, being, of necessity, accompanied by mineral gases which are exuded principally from the polar latitudes, comes in contact, on a clear and calm evening, with the vegetable gases in the lower atmosphere; either, then, a chemical action is induced between the two gases, causing intense friction and light; or the polarity of the mineral atoms in the upper atmosphere is deranged, and opacity or light is the result. We are inclined to think that both causes may be in operation at times, and thus we may account for the brilliancy and dullness of the displays—the bright ones being caused by chemical action. By the latter theory, the evanescent character and continuously shifting position of these northern lights may be accounted for more easily than by any other, for a shoot of the current along the "lines of force" could instantly reverse the poles of the atoms in the direction traversed; and in withdrawing as instantly allow them to resume their natural position again. The fact that nearly all these displays shoot out in lines from a Polar centre, very materially corroborates our assertion that the action of the "magnetic curves" of the earth is the principal cause of auroras.

The influence of the aurora on the compass needle and on telegraph lines, may then be easily accounted for by the Polar magnetic current.

But the auroras are seen much further south, how far south we believe has not been correctly ascertained; but, we should think, that under suitable conditions, they would be visible wherever any magnetic influence is found to be exerted on the compass.

It may be asked why do we not see the aurora to the south of us? We believe it is owing to the position of the atoms of the atmosphere, and to the atoms having two poles. The mineral atoms, under the influence of the polarity of the earth, being attracted southward through the oxygen, only those southern poles that come in contact with the oxygen exhibit light. Thus, also, for a somewhat similar reason, only those observers who are between the sun and a rainbow can see the latter. A rainbow cannot be seen if it exists between the point of observation and the sun.

Why is it only on calm and clear evenings that auroras are seen?

Because like dew, they are formed in consequence of a reciprocal action between the lower and upper atmospheres; a process which can only exhibit its action as the aurora, when the atmospheres are otherwise at rest, and when there are no winds to dissipate the gases, or divert them from the natural position necessary to form aurora; otherwise they form clouds, etc.

The sensation of cold is caused by the quantity of mineral gases mingling in the atmosphere.

The colours of the aurora are produced by the variety, and the different combinations, of the gases mingling; the vegetable gases causing the red, yellow and warm colours, and the mineral gases showing the blue and cold colours.

CHAPTER XXIX.

MEDICINE;
OR,
THE LIFE ACTION OF THE BODY, AND THE CAUSE AND CURE OF DISEASE.

A revolution in medicine.—The cause of disease unknown.—Incurable diseases.—Not creditable to the profession.—What are our bodies composed of?—What keeps up life in us?—What is blood?—How is blood formed?—How is the material we eat transformed into blood?—What causes and keeps up the circulation of the blood?—What is life?—The magnetic action of the body.—The function of the blood.—How the waste from the body is thrown off.—Hot water.—Purging.—Emetics.—The body compared to a fire.—Indigestion.—Consumption, its cause and cure.—Fevers.

IF the atomagnetic system be true, it is destined to revolutionize the whole practice of medicine. Any one who has been sick, must have noticed how ignorant certain doctors are as to the *nature* of disease, and how uncertain they are about the proper manner of treating it. The true cause of fevers and other diseases, the circulation of the blood, etc., is misapprehended, and in many cases totally unknown to them. Numerous diseases also are deemed incurable, such as consumption, heart disease, liver complaint, indigestion, and some others which ought to be no more so than measles, or scarlatina. It seems far from creditable to the profession that such should be the case, when, as a class, they have been operating on the human system for a term that is infinite, compared to the status of any other profession.

The name and purpose of every bone and muscle in the body is well known, but neither the true nature of the atomic material that composes the body, nor the nature of the action which keeps up life in it, is understood by them. The remedial knowledge of the profession, also, is obtained almost solely by an uncertain, or careless experience; and in many instances the remedy which has proved successful with one will be tried on another, without considering the differing nature of either case, relatively to the other.

How many cases of mysterious death also occur from the prescriptions of the physician, which are kindly ignored by the coroner, who is, fortunately, often one of themselves? Yet, notwithstanding all this, certain medical professors in Ontario, Canada, have petitioned their legislature for a bill, to constitute them the only life-preserving and death-dealing power in the land. To give them the power "to decide what is medical truth; to decide how truth shall be taught; to fix the standard of medical knowledge; and to prosecute all medical practitioners who may differ from them." Either the people in Ontario must be very simple, or their doctors are gifted with an audacity which could only result from ignorance. We hope for the honour and good name of the profession that there are no others in the world so bereft of reasoning capacity, for, it is evident, that a branch of any other profession or trade could as well claim similar privileges.

So much will, henceforth, be known about matter and its chemical properties, that disease should be, as Sir J. G. Simpson said, the exception, rather than the rule.

There ought to be no disease in existence but such as might be cured, if treated in time, as it usually exhibits its character early. A better knowledge of the qualities and character of medicine will also from hence be easily obtained—that is as regards its mineral or vegetable character—thus fewer mistakes should occur in transposing remedies, while experiments with new medicines might, with due care, be tried with impunity.

If we explain the life action of the body, it will show us what effect medicine is expected to produce and how it accomplishes its work.

In the first place of what are our bodies composed? Of matter in its varied forms and conditions; that is animal, vegetable and mineral atoms, in solid, liquid, and gaseous forms.

What keeps up life in the human system? Let us examine the human structure. There is a frame-work or skeleton of bones to support the body. There are muscles and nerves that work and guide the frame, and flesh and skin to protect these muscles and bones. But all throughout this flesh, muscle, and bone, there are multiplied arteries and veins filled with blood, coursing and saturating the whole, from the stomach to the extremities.

What is this blood?

If we bleed a man exhaustively, he becomes gradually weaker, till he has no strength to move, and finally he dies. He has lost his blood, and with that his breath of life; but all the rest of the machine, the bone, muscle, flesh, veins, etc., remain, yet they cannot keep life in the man; *his life support, therefore, must be through the blood.*

How is this blood formed?

If we take a strong, healthy, full-blooded man, and starve him for a time, we find he loses his flesh, and there is little or no blood or life left in him. Furnish him with warm water, he will revive, and may thus be kept alive for a time; but the reduced flesh is not restored. If we now take this famished and reduced man, and furnish him gradually with food and drink, he speedily becomes robust, fleshy, and full-blooded. The correct inference then is, that not only the blood, but the flesh, muscle, bones, hair, and everything connected with the body, must be formed from what we eat and drink.

How is this material that we eat transformed into blood? The stomach is analogous to a magnetic battery. The food which we eat is passed into it, and dissolved there by a process of chemical action between the particles of the animal, vegetable, and mineral matter contained in the food, similar to that which takes place in the mineral magnetic battery. This chemical action is produced through the agency of the water, or the fluids we drink, and with which the food is saturated. The portion which has dissolved, is then forced into the arteries and veins along with the blood, as new blood; while the more indigestible parts are ejected from the system, along with the waste material of the body, through the intestines.

But the blood itself is not life. It circulates through the body, we are told, once in twenty minutes. There must then be something to drive the new blood through the arteries and veins, in order to take the place of the old which has served its purpose, and is returning laden with impurities. For, while it is the function of the blood to restore the decayed parts of the body, it is also its function to carry off whatever inert material it wastes.

What then causes and keeps up the circulation of the blood?

We have already said the body, in its vital parts, is like a decomposing magnetic battery. In such a battery, the force produced by the dissolving of the minerals, is sent from the battery to the poles or extreme ends of the metal in connection. So is it with the body. While the food is being dissolved in the stomach, a force is generated which is continually urging the material, as soon as it is in a fit condition, from the stomach to the extremities. For the body is as much a magnet in its nature as a loadstone, although the magnetism is exhibited in a different way. This, therefore, is the primal force which keeps up the circulation of the blood. *Life then is the atomagnetic action of the body.*

In order now to discover the cause of fever and disease in the system, let us examine again the function of the blood, and the action of the body in throwing off its waste material, and of rebuilding the same. The waste is thrown off in two ways; firstly, by the exhalations from the body externally all over its surface; and secondly, by passing along with the blood to the intestines, and being ejected from the body with the insoluble parts of the food eaten. It will be seen, therefore, that should there be a checking of the circulation in any way, as by cold or chills, or by the food not being digested as well, or as readily, as it ought to be, or by the waste material clogging up the veins, or by uncleanliness checking and filling up the pores of the skin, some disease or ailment is sure to ensue. In a stoppage of this sort, the atoms of the body interfere with, or obstruct, the general arrangement of its working, and a new *local* action takes place in the part so clogged, causing inflammation, fever, palpitation, rheumatism, cholera, bodily pain, or a condition and disposition to take on disease in the form of any contagion or infection that may be prevalent.

As we cannot always foresee, and so prevent a stoppage of the bodily functions, no household should be without the knowledge and means of restoring the circulation of the blood.

The first remedy on the list we commend is hot water. Inasmuch as the blood and other material of the body is largely composed of water, should any poisonous matter exist in the human system that is foreign to it, the hot water will dilute it, and assist in urging it along to the intestines and extremities to eject it from the system. In many cases of lassitude or exhaustion, no other remedial agent would be required.

The next agent, is a purgative food or medicine adapted to the chemical condition of the body; and in urgent cases, an emetic—adapted to the same conditions—given in a weak solution, in doses every twenty minutes, or at sufficient intervals to allow it to permeate the whole body through the circulation of the blood, before vomiting takes place. This, together with a moderate application of moisture and heat to the body, would generally be a certain cure. Some simple treatment such as we have described, administered in time, would in many instances check what might otherwise result— by the prescribed system of treatment—in a long and serious illness.

As regards the circulation of the human body, it may be compared to a fire. If the fire is expected to burn well, there must be freedom for the air to circulate through the material that feeds it, and therefore it wants occasional clearing of the ashes, before having a fresh supply of fuel put on. Should the ashes be allowed to accumulate, the fire will burn low and threaten to go out; then the extensive poking which is necessary to restore a draft, almost puts the fire out. So it is with the body if the circulation is neglected, the body becomes feeble and the spirits languish, then sickness, cold or fever ensues, and unless the circulation is speedily restored in a proper manner, the life of the patient is apt to go out.

Indigestion is one of the most common complaints, and the cause of many ulterior diseases. It is caused, in many instances, by an insufficiency of liquid to dissolve the food and to enable the body to keep up a free circulation. In England, where great quantities of beer and wines are drank during dinner, indigestion is not half so frequent as in America, where iced water is the too frequent accompaniment to every meal. Our food and drink should always be taken warm when possible, because a certain temperature must be kept up in the body to digest the food. Sufficient salt, too, should always be taken at every meal— especially in warm weather, and in equatorial latitudes.

What causes consumption? Inflammation. What causes inflammation? Merely the circumstance of certain veins or arteries becoming clogged with waste material. This is caused by a chill in the region of the lungs, tending to check the circulation; or may arise from neglect or improper care of the body. A local action then takes place, and a pain is felt in one of the lungs. The doctor probably explains that the lungs are wasted or decayed in that particular spot, and that the only treatment he can recommend is a change of climate, or a *blister*. If the patients do not go away from home,—and there are comparatively few who can afford to do so—then they may make up their minds to die. It seems like murder that some hundreds or thousands of people should die every year of consumption, and that the profession which is so greatly accountable for the calamity should never have found a remedy for the disease. Those who try to discover a cure also, are treated discourteously instead of offered assistance.

How would we treat consumption ?

Not so much by drugs, or blisters, as by careful nursing. From what we have said of the nature of the body and its mode of action, it may be inferred that we would in the first place restore it to a proper working condition, by attending to the digestion and the circulation, so that the body be kept warm in every part, by its own natural atomic action. We would then find the locality of the inflammation, and by the continuous application of moisture and heat, along with a slightly mineral solution to the part affected, we would expect to produce a reciprocal action between the atoms in the region of inflammation, and in the liniment, which would neutralize each other, and so induce a healthy circulation, that might result in removing the state of inflammation entirely. If carelessness, or inattention to the rules of health should bring on the inflammation again, the same treatment would have to be repeated. Thus no young or otherwise healthy person need die of consumption if properly prescribed for. It will be evident, therefore, that all those compound nostrums, patent medicines, and galvanic batteries, which are advertised indiscriminately to cure consumption, without some general treatment of the body and system besides that of local application to the parts affected, are a deception, and destined only to failure.

How are fevers caused and cured?

Fevers, as we have already said, are caused by the checking, of the circulation, owing to the blood being vitiated. To bleed, in such cases, may for the moment relieve by reducing the *quantity*, but certainly can never alter the *quality* or condition of the blood. The dogma that every fever must run its course of so many days, and that its various changes and stages should be watched and named, is a most absurd and dangerous one; for, admitted that the fever is caused by vitiated blood, the common sense treatment would obviously be to chemically change its condition, and to induce a free and full circulation of the blood throughout the body: not to wait till symptoms of some particular type of fever or other disease be developed, and the whole bodily system is reduced and paralyzed. Or worse still, ere symptoms are fully developed, to guess at the disease and prescribe a strong medicine, which, if the surmise should be wrong, may probably prove to be the worst drug the patient could have taken.

Suppose two men are in a room with a fever patient, and both are attacked by the fever, but one man recovers in a few days while the other has the fever for weeks; it will then be commonly said that the one who recovered did not have the fever at all. Is it not more likely that the constitution or condition of the one was able to resist, or counteract the influence of the disease, and thus throw it off sooner, while the other wanted assistance from a physician, but the proper means were not used. It is only reasonable to suppose, that if a strong healthy man is able naturally or voluntarily to throw off a fever, a weak man with proper assistance and medicine ought to be able to do so too.

It is not our province, in a work of this kind, to enter into a detailed system of medical treatment, for such a proceeding would enlarge the volume into a library; but we desire principally to call attention to the fact that many diseases are deemed incurable, which properly ought not to be so called; and also that the few illustrations we have given of the cause of disease and the manner of treating sickness, may lead those who feel sick to trust to and to act in the first place for themselves, until a better system of medical treatment is adopted; thus many valuable lives may be saved every year.

CHAPTER XXX.

ATOMAGNETISM AND RELIGION.

Religion not affected by Atomagnetism.—The inherent life in atoms, and the spontaneous development of the mind, seem grand arguments for the Materialist. — The movements of Planets and Comets.—The great machinery of the universe.—What need of a God ?—Man fancies himself a Monarch.—No animal intelligence his superior.—Only a parasite.—Chained to the earth.—On a level with his dog.—Matter existed without properties.—Who endowed it with them ?—Divine mind of man.—Magnetism not God.—How simple, miracles must be to Him who formed and holds the key of natural law.—Insignificance of man.

As Religion may be considered to be affected by the system of science promulgated in the previous chapters, we have reserved our remarks on the connection between them for a concluding chapter.

It may seem, to many readers, that we supply arguments from which the materialist could still further advance his cause. When we say that matter is possessed of inherent life, and that this life is capable of forming any production on earth spontaneously, according to the nature, condition and position of the materials commingling, not only in the mineral and vegetable kingdom, but in the animal kingdom as well ; it appears a most astounding assertion. But when we state that the same atomic matter spontaneously produces the mind, not only in the unintellectual animal but in man, and that those wonderful instinctive powers which seem to baffle the wisdom of man, in the bee and the beaver, are only manifestations of the same power inherent in the atoms of matter; might not the infidel exclaim, what more do we want? Furthermore, if the power of matter controls and governs the universe, guides the planets in their course, steers the comets in their erratic wanderings through space, and shields those bodies which come in their path; if the mysterious meteoric shower, the awe inspiring thunder, the earthquake, and every movement in earth, sea and atmosphere, from the tiniest individual atom to the bright orb above us, are all controlled by the same inanimate force, so that each is powerless to work except as that law impels it, and each moves but as a wheel in the great machinery of the universe! might not the materialist exclaim; what need of a God at all? But, here insignificant man fancies himself monarch of all he surveys. His lofty mind looking down on the brute creation, and seeing nothing which could be deemed his superior, makes a God of himself. He forgets that this earth is infinitely less than millions of others in space. He sees those brilliant worlds above him, but fancies they are only jewels studded in the skies, to add to the glories of his earthly home. He forgets that they are peopled with beings perhaps many times more able than himself, and with intellect as much above his, as his own is above that of his dog! He forgets that he is merely a parasite of the earth, chained to its surface without hope of escape, while those other beings may be gifted with angels wings to soar from star to star. He forgets that the dog he spurns from his feet is made of-dust like to himself, and yet he would place himself on a level with Him who created the heavens and the earth, and caused both himself and his dog to grow upon it.

While we deny that our system favours the materialist or the infidel, we maintain that it brings the most conclusive evidence to bear against them. In the science of the day mystery is found on every hand, and no two observers read phenomena alike. A man may in consequence be excused for holding opinions of his own on science and religion, when his statements cannot be disproved. But, as we have laid down the law which regulates all nature, which originates and controls all forces in it, and is therefore as infallible as nature itself—the possibility of any misconception or misconstruction is obviated entirely. The existence of the Great God, the Designer and Ruler of all, is thus seen at once to be true, for in the perfection of the law, is shown the perfection of Him who created it.

While then we say that matter has inherent properties, we have also said that it may exist in a position not to exhibit its properties; as in the beginning of our earth, when, as we are informed, the material forming the earth, sea, and atmosphere, was "chaos" or invisible matter—"Without form and void." Grant, therefore, that matter did exist without properties; where is the materialist who will explain how the properties were imparted to it without a God? The fact, too, that man has a divine mental faculty or mind, distinct from his animal or earthly mind, is also conclusive evidence of a power beyond that contained in matter. We have ourselves been told that because we stated life to be magnetism, we therefore implied that magnetism was God. But if magnetism is a power with which matter was endowed, there must also of necessity have been an endower of that power.

This latter fact strikes a vein of thought, which convinces us of the illimitable power of the Creator.

If atomagnetism so regulates all atoms and forces of matter, that, not the slightest movement, not the waving of a blade of grass, not the flutter of a bird on the wing, not the fleeting of a cloud in the sky, but is governed by this law; then how great is the power of Him who holds the key thereof! He who gives a power can surely take it away again; and He who can create one power, can surely devise another. How simple a matter then must it be to the Deity to shut the mouths of lions; to take the heat from a burning fiery furnace; to part asunder the Red Sea; and to raise Lazarus from the dead. The fact that not only these mysteries, but the creation of our world, the flood which deluged it, and all other natural phenomena recorded in Scripture are found by this atomagnetic system to harmonise with scientific facts; and the impressive and yet simple fact that the infidel's own lips but open and shut, according to the same law, should show proud man what a miserable and insignificant being he is, and how much he is at the mercy of an ALMIGHTY RULER.

INDEX.

A.

Adam, blamed for all the ills that afflict mankind, ... 109
African, Atmosphere of the ... 125
Agassiz, Louis, on living beings born of eggs ... 21
Agassiz, Louis, on an animal and divine mind ... 36
Agassiz, Louis, on men and monkeys ... 22
Agassiz, Louis, on the creation of the many at the beginning ... 23
Agassiz, Louis, on special creation ... 27
Agassiz, Louis, on coral insects... 148
Agassiz, Louis, on millions of coral insects fed from one mouth ... 148
Agassiz, Louis, on coral growing by budding ... 149
Agassiz's, Louis, argument against Darwin with coral insects ... 149
Agassiz, Louis, on the age of Florida reefs ... 149
Air, no moisture in ... 47
" no, in water ... 115
" no, in ice ... 116
" Mr. H. Higgins on water absorbing ... 116
" composition of ... 124
Albion Mines, Nova Scotia, site of a volcano ... 153
Alphabet, similarity between the elements of matter and the ... 5
America, rain diminishing in North ... 114
" rain increasing in Spanish ... 114
" disastrous effects of clearing the forests in North ... 114
" ruined cities in Central ... 115
American, weather maps ... 132
Animals, possessed of one mind only ... 32
" have no soul ... 32

Animals, their mind a property of earthly matter only ... 32
" how produced spontaneously ... 21
Appetite is incipient mind ... 30
" the, of spontaneous insects ... 30
" what is ... 30
" governs all animals up to man ... 31
" a seal's ... 31
" a calf's ... 31
" how to acquire an ... 32
Atomagnetism, the law of matter and its force ... 6
Atomagnetism, how great the power of Him who holds the key of ... 199
Atomagnetism supersedes gravitation ... 93
Atomagnetism, the motive power of the whole machinery of the universe ... 197
Atomagnetism, Religion, and... 196
Atomagnetism, explains all mysteries in Scripture ... 199
Atomagnetism brings the most conclusive evidence to bear against infidels ... 198
Atoms, their character ... 1
" two great classes of ... 2
" Mineral and vegetable... 2
" male and female ... 2
" properties of ... 2
" all alive ... 3
" are all magnets having polarity ... 7
" the attraction of like, explained ... 7
Atomic action, continuity of... 123
" likened to a gossamer thread ... 59
" matter, spontaneously produces mind ... 196
Atmosphere, the, in layers ... 106
" temperature of, in balloon researches 125
" composition of ... 125

Atmosphere, everything has an 125
" of the African 125
" extent of the planets 163
" pressure of, causes tides 164
" pressure on, causes meteors 130
Atmospheric circulation, theory of 171
Atlantic, S. S., loss of 101
Attraction not the cause of the tides 162
Artesian wells for the prevention of earthquakes 159
Auroras, facts concerning 182
" Mairan on 182
" Lardner on 182
" M. Biot on 182
" disturb the magnetic needle 183
" caused by the magnetism of the earth 183
" two theories of 183
" colours of, caused by 185

B.

Balloon researches, by Guy Lussac 125
Bastian, Mr. H. Charlton on Origin of lowest organisms 28
Battles, followed by rain, cause of 117
Beasts have only one mind 34
" have no soul 35
" know not sin, nor do wrong 35
" have appetites but no desires 35
Beavers, foreknowledge of 34
Bees, " " 34
Beche, Sir Henry de la, on carbon 141
Beer, vast quantities of, drank in England 192
Berkeley, Bishop, on matter 1
Biot, M., on auroras 182
Bible, the, comprehensiveness of 14
" proof of vegetable life drawn from 14
Blood, what is 188
" function of the 189
" cause of the circulation of the 190
Botanical theories of vegetable life 14

Botanical theories of trees 149
Bottled sunshine 81
Borer, the, worm holes in coral made by 147
Body, Human, a machine easily injured 133
Brain, the, a picture gallery 41
Branches and leaves, why they spread 16
Brewer, Dr., on the sun the source of heat 56
Buckland, Frank, on a chicken's instinct 33
Buoys, scientific, for leading men astray 81

C.

Calms, cause of equatorial 129
" cause of, in the centre of cyclones 131
Cambodia, ruined cities of 115
Cambridge, Lectures, Agassiz 148
Carbon, trees composed of 140
" Sir Henry de la Beche on 141
" said to be injurious to man 141
" eaten by man 141
Carpenter, Dr. W. B., his theory of oceanic circulation 168
Carpenter, Dr. W. B., his deep sea soundings 168
Carpenter's, Dr. W. B., experiment with glass trough 168
Carpenter, Dr. W. B., Dr. Wyville Thompson against 169
Carpenter, atmospheric and oceanic theory overlooked by 171
Cells in plants, what impels them to divide 16
Central America, ruined cities in 115
Celestial Dynamics, Dr. Mayer's 154
Chamber's Journal on Dew 119
Charcoal in the Nova Scotian coal fields 142
Challenger, H. M. S. 146
Chamber's Encyclopedia on coral insects 146
Chamber's Encyclopedia on the time taken to fill a gap in a coral reef 152
Cheese, life in 20
" Chinese 136
Chemical action, Prof. Grove on 44
" all changes caused by 44
" Digestion caused by 44

INDEX.

Chemical action the great destroyer ... 45
" " examples of... 45
Child, Dr., on spontaneous generation ... 24
Chicago, once a fever swamp... 115
" snow storm at ... 132
Chinese cheese, made of peas and beans ... 136
City, a, why it is free of fever and ague ... 115
" of Washington, S. S., loss of ... 101
Cincinnati, once a fever swamp 115
Circulation of the blood, cause of the ... 199
" of the blood in the body compared to a fire ... 191
Clouds, formation of 110, 123
Common sense, Sir John Herschell on ... 109
Common sense a poor guide to the study of science 109
Coal, how formed ... 56
" of vegetable origin ... 139
" Prof. Rogers on how it is baked ... 140
" formed by petrifaction... 143
" Nova Scotia, fields ... 142
" inexhaustible ... 144
" a forest will only produce half an inch of ... 140
" how hydrogen entered the 142
Cochineal, a parasite of the cactus ... 147
Cobalt blue, magnetic ... 83
Columbus, Christopher ... 53
Colour, length of red and violet rays of light ... 77
" vibratory theory of ... 79
" Herschell, Tyndall, Maxwell, Helmholtz on 80
" Atomagnetic theory of... 81
" a property of matter... 91
Compass magnetic, exhibition of mineral life ... 10
" cause of deviation of, in ships ... 98
" general ignorance of its action ... 99
" influence of a cargo of petroleum on the... 99
" disturbed by auroras, why ... 183
Comets, two or three visit the earth every year ... 173
" facts connected with ... 173
" not absorbed by the sun 173

Comets, Jupiter attracts ... 174
" periodical, Halley's & Encke's ... 174
" some have not tails ... 174
" of 1680 ... 174
" why repelled from the sun ... 175
" how made periodical ... 175
" have atmospheres ... 176
" cause of the luminous tails of ... 176
Consumption, cause of ... 192
" the doctors on ... 192
" cure of ... 193
" no young person need die of ... 193
Cornhill Magazine on seed bearing meteors ... 13
Coral insects, sermons preached on their industry 145
" " unworthy of notice 145
" " live in shallow water ... 146
" " merely parasites... 147
" " Agassiz on millions of, fed from one mouth ... 146
" " Agassiz's argument against Darwin with ... 149
" " Darwin's theories of 149
" " Sir John Herschell on ... 150
" " how they keep the Coral Islands above water ... 16
" " a species of animal jelly ... 148
" " artistic taste of... 147
" " possessed of administrative and governing powers... 152
Coral, how it grows ... 145
" found only in equatorial latitudes ... 145
" grows by budding ... 149
" Florida reefs of ... 149
" what causes the colour of 148
" Islands, said to be sinking 150
" " not formed wholly of coral ... 151
Coroner, danger of having a doctor as a ... 186
Cow, is a, the mother of maggots? ... 20
Crosse, Mr., producing insects spontaneously ... 20
Creator, the, illimitable power of 199

Crystallization, coral a form of 147
" a form of mineral life ... 11

D.

Darwin, Dr. on one primordial form ... 22
Darwin's, Dr., theory of the origin of species overthrown... 15
Darwin, Dr., on mind ... 30
" " on the power of wishing ... 28
Darwin, Dr. Agassiz's argument with coral insects against ... 149
Darwin's, Dr., theories of coral insects ... 149
Darwin, Dr., states the Coral Islands are sinking ... 150
Darwin, Dr., quoted by Sir John Herschell ... 150
Daylight, cause of ... 75
Descartes on matter and motion 6
Davy, Sir Humphrey, on heat in ice ... 58
Deep sea, the, temperature of... 168
" " animal life in ... 168
" " coldness of the, accounted for ... 172
" " currents, Dr. Carpenter's theory of ... 165
" " Dr. Wyville Thompson on ... 169
Dew, Chamber's Journal on ... 119
" Baptista Porta nearly discovered the true theory of 119
" Aristotle on ... 120
" Muschenbrook delayed the discovery of the true theory of ... 120
" Dr. Wells discoverer of the accepted theory of... 121
" produced similarly to water ... 123
" a calm and clear evening essential for the formation of ... 123
De la Rive, on two differing electricities ... 86
Delusions, scientific, about magnetism ... 94
Deviation of ships' compasses, caused by a speaking trumpet 100
Deviation of ships' compasses, caused by steering apparatus, 101
Deviation of ships' compasses, caused by a cargo of petroleum 99

Deviation of ships' compasses, caused by iron cross trees ... 101
Diseases, doctors ignorant of the nature of ... 186
Diseases, so-called incurable ... 186
" should be the exception, as Sir J. Y. Simpson said, ... 187
" the cause of ... 190
Distillation, the common theory of rain a process of ... 109
Digestion, what is ... 138
" extraordinary case of, by coral insects ... 148
Doctors, ignorant of disease ... 186
" their ignorance not creditable to them 186
" life action of the body unknown to ... 187
" their remedial knowledge gained only by experience ... 187
" mysterious deaths by, ignored by the coroner 187
" audacity of, in Ontario 187
" one class of, petitioning for a bill to persecute others ... 187
Dogfish has no gills ... 116
Drainage, Sir John Herschell says, lessens the rainfall ... 114
Dumas, M., on origin of life ... 20
Dunsappie Loch, near Edinburgh, Scotland, obscure ice on ... 112
Dust and Disease, Tyndall on... 26

E.

Earth, the, heat generated by, falling into the sun 61
" " a magnetic battery 73
" " no colour on ... 80
" " Prof Rogers on the central heat of 140, 153
" " thickness of the crust of ... 153
" " Dr. Mayer on the internal fire of ... 154
" " Drs. Tyndall, Thomson, and Herschell on ... 154
Earthquakes, Sir John Herschell on 155, 159
" how prevented ... 159
" caused by chemical action ... 158
Ecliptic, plane of the, tides caused by pressure on ... 164

INDEX.

Eggs, not necessary for producing life ... 21
" Agassiz on ... 21
" not chickens ... 87
Egypt, rain becoming more frequent in ... 114
" cultivation of the palm in 114
Electricity, Dr. Thos. Thomson on ... 85
Electricity, Parker, Sir Wm. Thomson, Tyndall, Grove, and De la Rive on ... 85, 86
Electricity confounded with magnetism ... 84
Electricity not a force, only combustion ... 87
Electricity caused by magnetism ... 87
Electricity, all light is ... 88
" lightning is ... 89
Emission theory of Light, Sir Isaac Newton's ... 66
Emetic, efficacy of an, in sickness 191
Encke's comet ... 174, 175
Energy, Professors Thomson and Tait on ... 69
English navvies, the work they do ... 137
Equator, calms of the, explained 129
Esquimaux, Sir John Ross on the appetite of ... 137
Evaporation, a chemical action.. 110
" induced by cold as well as by heat... 110
Evolution, doctrine of ... 27
Experiments, laboratory, doubtful helps to science ... 58
Experiments, Nature's, to be studied ... 59

F.

Faith, essential in religion ... 130
Faraday, Michael, on magnetism 91
" " on magnetic curves ... 92
" " born too soon 92
" " led astray by his imagination regarding magnetic curves ... 130
Faraday, Michael, how he led Maury astray ... 130
Fevers how caused and cured ... 193
Fire mist ... 4
Fireflies, light of ... 64
Filings, iron, magnetic arch formed by ... 93
Fire, not so powerful as water... 107

Fish produced spontaneously ... 27
Fishes gills used for filtering their food, not for breathing.. 116
Florida reefs, Agassiz on the age of ... 149
Food, Prof. Lyon Playfair on ... 134
" Liebig on ... 137
" Playfair's tables of ... 134
" on mineral ... 137
" how transformed into blood ... 180
Fogs, cause of 110, 111, 123
Forests, why they produce rain.. 114
Forms of water, Tyndall's ... 109
Fox, sunning of the ... 35
Fraser's magazine on matter ... 4
Frost leaves ... 93
" hoar, cause of ... 123
Fundy, Bay of, North America, high tides in, cause of 161, 165
Fundy, Bay of, rise of two feet in three miles ... 166

G.

Gastric Juice, what is ... 138
Germs, theory of ... 25
" difference between atoms and ... 25
" where do they come from 25
" must be born in some way ... 25
" Pasteur & Tyndall on 25, 26
Geography of the sea, Maury's.. 130
Geysers and hot springs, cause of 54
" Great, California, 54, 158
Ghoorkas, flesh eating ... 137
Gills of a fish used for filtering food, not breathing ... 116
Gibraltar, upper and under current at Straits of ... 165
Glaciers, why chasms in Alpine, are blue ... 83
" Tyndall on solar heat the cause of ... 109
God, magnetism not ... 193
" perfection of ... 198
" the power of ... 199
Gossamer thread, atomic action likened to a ... 59
Graham, Prof., on the Leonarto meteorite ... 180
Granite, water in ... 109
Gravitation, the law of, upset... 98
" fails to account for the repelling power of the sun ... 174

Gravitation as explained at the present time ... 175
Grass, why green ... 81
" why dew deposits on ... 121
Grindon, Leo. H. on life ... 22
Grove, Prof., on matter ... 1
" confesses ignorance of the source of heat ... 61
" Prof., on electricity ... 86
" on magnetism ... 96
Gulf Stream, no longer a mystery 167
" Dr. Carpenter on... 167

H.

Halley's Comet ... 174, 175
Hens, how they transmit their mind to their chickens ... 35
Heat, natural ... 53
" by combustion ... 53
" by friction ... 54
" Dynamical theory of ... 52
" Tyndall on ... 52, 60, 61
" Mayer on ... 53
" as a mode of motion ... 58
" Grove, Lardner and Brewer on ... 61
" independent of matter ... 61
" Tyndall on sound generating ... 102
Hegel on mind ... 37
Helmholtz on colour ... 80
Herschell, Sir J. F. W., on the waste heat of the sun ... 71
Herschell, Sir J. F. W., on the temperature of space ... 71
Herschell, Sir J. F. W., on colour ... 80
Herschell, Sir J. F. W., on common sense ... 109
Herschell, Sir J. F. W., on the formation of rain ... 109
Herschell, Sir J. F. W., opposed to Kamtz's theory of rain ... 113
Herschell, Sir J. F. W., on altering the weather ... 113
Herschell, Sir J. F. W., on the reason why trees attract rain.. 114
Herschell, Sir J. F. W., on drainage being bad ... 114
Herschell, Sir J. F. W., on weather and weather prophets 128
Herschell, Sir J. F. W., on the cause of wind ... 128
Herschell, Sir J. F. W., on coral insects ... 150
Herschell, Sir J. F. W., on volcanoes ... 155
Herschell, Sir J. F. W., on earthquakes ... 159
Herschell, Sir J. F. W., knew of chemical action in the interior of the earth ... 159
Herschell, Sir J. F. W., his failure ... 160
Herschell, Sir J. F. W., on comet of 1680 ... 174
Hercules, collision with ... 46
Hills, why purple ... 83
Horse latitudes, calms of the ... 129
Holland, criminals fed on food free from salt in ... 137
Hound, scent of a ... 34
Hope, on Origin and Prospects of man ... 37
" on mind ... 37
Hot Springs and Geysers, cause of ... 50
Huxley, Prof., on matter and motion ... 6
" " on origin of life 28
Hydrogen, in coal ... 141
" how introduced into coal ... 142
" All mineral matter is ... 2
" in the Leonarto meteorite ... 180

I.

Icarus, fate of ... 28
Ice, action of magnetism in shell 93
" hoar frost, spears of ... 123
" composition of ... 172
Icebergs, why granulated ... 172
Inertia, action of, on the tides... 165
Indigestion, The Scotch comparatively free from ... 188
Indigestion, cause of ... 192
" not so common in England as in America ... 192
Indian, animal mind of an ... 39
Instinct, a higher phase of mind than appetite ... 33
Intellectual Observer on rain ... 112
Iron, why rust reddens ... 84
" cross trees, dangerous nature of ... 101
Iron ships, great mortality in... 99
" great care should be taken in fitting up... 99
" the steering apparatus in, dangerous ... 101
" the iron stanchions and davits of, are magnetic poles ... 100

J.

Jupiter, the time light takes to come from ... 67

K.

Kamtz on rain ... 112
" his theory opposed by Sir John Herschell... 113
Kepler as an astronomer ... 69

L.

Lamarck, M., on origin of life... 25
" on monads ... 25
" Sir Chas. Lyell on 25
Language and matter, analogy between ... 5
Laplace, on the revolution of the earth ... 154
Lardner, Dr., on heat independent of matter ... 56
Lardner, Dr., on the moon causing the tides ... 162
Lardner, Dr. on Auroras ... 182
Lead tree, example of mineral life ... 11
" tree, caused by magnetism 96
Leonarto meteorite, Prof. Graham on ... 180
Life, the atomagnetic action of the body is ... 190
" mineral ... 9
" vegetable ... 11
" magnetism is ... 193
Light, Sir Isaac Newton's emission theory of ... 66
" undulatory theory of ... 66
" Tyndall's waves of ... 67
" rate of travelling ... 67
" how many waves of, make an inch ... 67
" how many red and violet waves of, enter the eye in a second ... 67
" mistake made between sight and ... 67
" candle experiment ... 67
" cannot travel ... 67
" of stars travelling to us all nonsense ... 68
" phosphorescent, on the Pacific ... 63
" cause of ... 64
" of fireflies ... 64
Locke on matter ... 1
Lobster has no gills ... 117
London air, Tyndall on ... 26
Luminiferous ether ... 93
Lussac Guy, balloon researches by ... 125
Lyell, Sir Charles, on Lamarck's monads ... 25

M.

Macartney, discovering fish produced spontaneously ... 27
Magnets how to make them without hammering ... 94
Magnets, absurdity of fixed, in ships ... 99
Magnets, all heavenly bodies are 175
Magnetic Battery, how arranged 75
" " the stomach like a ... 189
" curves, the cause of... 92
" " Faraday unable to explain them ... 92
" " Tyndall on ... 93
" " the cause of storms ... 130
Magnetic curves, Faraday's fallacies regarding ... 130
Magnetic curves, Maury led astray by Faraday's theories of 130
Magnetic curves, cause of auroras ... 183
Magnetism, the universal force in nature ... 90
Magnetism, the cause of germs.. 24
" Prof. Grove on ... 90
" Faraday on ... 91
" an inherent property of matter ... 91
" of iron railings ... 94
" of iron ships ... 100
" changes with position ... 94
" in straight lines ... 93
" not God ... 198
" matter endowed with 198
Mairan, on auroras, and extent of sun's atmosphere ... 182
Mallet, Prof., on volcanoes ... 153
Man, animal and spiritual mind of ... 34
" pride of intellectual ... 197
" a parasite of the earth ... 199
" on a level with his dog ... 197
" insignificance of ... 199
Matter, Grove. Locke, Berkeley, Tyndall, Norton on 1, 2, 3
Matter, eternal and resolvable into atoms ... 2

Matter composed of male and female atoms ... 2
Matter, Vestiges of Creation on 3
" Park's Chemical Catechism on ... 4
" analogy between the alphabet and ... 5
" without properties ... 37
" and its force ... 6
" and motion, Huxley, Descartes and Newton on ... 6
" and motion a senseless dance of the atoms ... 6
Maury, estimate of ... 130
" led astray by Faraday... 130
" on circular storms ... 130
Maxwell, Prof. on colour ... 50
Mayer, Dr., on motion ... 53
" on the internal fire of the earth ... 154
" on volcanoes ... 155
" his doggedness ... 155
Medicine, whole practice of, changed by atomagnetism ... 186
Mediterranean, rise of the tide in the ... 161
Mediterranean, cause of the low tide in the ... 165
Mediterranean, the under current in, tested in 1830 ... 165
Memory, where located ... 41
Meteors, Sir Wm. Thomson on seed bearing 12, 178
Meteors, theories regarding ... 178
" facts concerning ... 178
" no ring of ... 179
" Prof. Newton on the inflammable nature of ... 179
Meteors, Sir Wm. Thomson on the probability of, colliding 179
Meteors, atomagnetic theory of 180
" caused by pressure of the atmosphere ... 180
" Schiaparelli on comets causing ... 181
" November showers of... 181
Meteorites, Dr. Sorby on the structure of ... 180
Meteorite, Prof. Graham on the Leonarto ... 180
Mexico, phosphorescent waves off the coast of ... 176
Milk, life in ... 21
Mind, Schelling, Hegel, and Hope, on ... 37

Mind, Agassiz on ... 38
" situation of the ... 41
" what composed of ... 40
Mineral life, lowest form of life 9
" " infallibly correct... 9
" " the compass needle 9
" " philosopher's tree an example of ... 10
" " coral and crystallized candy examples of ... 11
Monads, Lamarck on ... 25
Mount St. Helena, California, an extinct volcano 51, 55
Moon, the, light of ... 75
" tides high at new and full ... 161
" how it influences the tides ... 164
" Lardner on the influence of the, on tides ... 162
Moonblindness, Dr. Wells on the cause of ... 121
Musohenbrook on Dew ... 120
" stupidity of ... 120
Muskrat, foreknowledge of the... 34

N.

Napa Valley, California ... 55
Navvies, English, work done by 132
Newton, Sir Isaac, on matter and motion ... 6
Newton, Sir Isaac, on the emission theory of light ... 66
Newton, Sir Isaac, on the apple falling ... 97
Newton, Sir Isaac, his law of gravitation upset ... 98
Newton, Sir Isaac, not satisfied with his discoveries ... 98
Newton, Sir Isaac, on comet of 1680 ... 174
Newton, Prof., Yale College, America, on meteorites ... 179
Nicaragua, Lake, effect of rainfall on ... 114
Nitrogen, classification of ... 124
Norton, Prof. W. A., on matter 3
North America, the rainfall diminishing over ... 113
Nova Scotia, dangerous coast of 101
" " transparent ice on Pictou River ... 112
" " coal fields of ... 142
" " Albion Mines in... 158

O.

Oceanic circulation, Dr. Carpenter's theory of ... 167
" Carpenter's experiment to prove ... 168
" Dr. Wyville Thompson against Carpenter's theory of ... 169
" No necessity for an ... 169
Oil boring in Pennsylvania ... 159
Opacity, cause of ... 111
" caused by combination of gases ... 132
" cause of auroras' light 184
Origin of life ... 18
" " Sir Wm. Thomson on ... 12
Origin of lowest organisms, by H. Charlton Bastian ... 26
Ozone, the discovery of, nonsense ... 127
" contradictions regarding 127
" matches generate ... 127

P.

Palenque, Central America ... 115
Parker's School Book of Philosophy on electricity ... 85
Pasteur, M., on spontaneous generation 24
" on germs ... 24
" on preseved meats 25
Panama, Isthmus of, coral on the ... 146
Park's Chemical Catechism on matter ... 4
Parasites, all plants and animals 23
" of fruit, raspberries, figs, etc. ... 23
Pennsylvania, oil boring in ... 159
Petrifaction, coal made by ... 143
Pictou River, Nova Scotia, transparent ice on ... 112
Pigeons, instinct of ... 34
Philosopher's tree ... 9
Playfair, Prof. Lyon, on food... 134
" supports Liebig's division of food ... 134
" food tables ... 135
" on the close connection between animal and vegetable substances ... 136
" on English navvies and Arabs ... 137

Playfair, knows not the action of the mineral part of food ... 137
Phosphorescent waves off the coast of Mexico ... 176
Plants, do not breathe, but exhale ... 16
" absorb sustenance from the earth ... 16
" result of painting ... 16
Poles, every atom has two 7
" like, repel; unlike, attract 7
" North and South, why cold ... 172
Polarity of atoms ... 7
Polypes, coral insects, Agassiz on ... 148
" Chamber's Encyclopedia on ... 146
Porta, Baptista, nearly discovered the true theory of dew ... 119
" denied the moon and stars caused dew ... 120
" thought dew was condensed from air ... 120
" shewed that dew rose from the earth ... 120
Prairies in coal fields ... 142
Prescott's History of the Electric Telegraph ... 86
Preserved meats, why they decompose ... 25
Pressure on the atmosphere, the cause of tides ... 162
" The cause of meteors ... 130
Proctor, Prof. R. A., on rain 112, 113

R.

Rainbow, why not seen between the observer and the sun ... 185
Rain, how accounted for ... 109
" assists in the formation of glaciers ... 109
" Sir John Herschell on ... 109
" does not come from the Gulf Stream ... 111
" formed in the atmosphere around us ... 111
" does not necessarily fall a great distance ... 111
" Prof. Proctor on ... 112
" Kamtz on ... 112
" shot out of a cloud ... 113
" becoming more frequent in Egypt ... 114
" why trees attract ... 114

Rain, why forests produce ... 114
" great battles followed by 117
Rainfall, diminishing in North
    America ... 114
" Herschell says drainage lessens the ... 114
Rain guage, the nearer the ground the greater the rainfall indicated ... 111
Reefs, coral, Agassiz on the age of Florida ... 159
" Darwin's theories of ... 150
" how formed ... 151
" how a gap was filled up ... 151
Religion, faith essential in ... 130
" and atomagnetism ... 196
Renaissance, Painting and Architecture ... 60
Rogers, Prof., on winds ... 128
" on coal and petroleum ... 140
" on coal being baked ... 140
" on hydrogen in coal ... 141
" on the earth's internal fire ... 140
Roots, why they spread ... 16
Ross, Sir John, on the appetite of an Esquimaux ... 137
Ruskin, John, his crusade against Renaissance art ... 60

S.

Sand bars, how caused ... 151
Salt, used in great quantities by Africans and Arabs ... 137
" Canadian Indians eat no, in winter ... 137
" criminals fed on food free from ... 137
" action of, in the body ... 138
" necessary for digestion ... 192
Sable Island, sand bars of ... 151
Science, absurdities of ... 78
" looseness in ... 78
" like a voyage of discovery ... 60
" convention in Portland, U. S., 1873 ... 78
Scientific use of the Imagination, Prof. Tyndall's ... 82
Schelling on mind ... 37
Scriptures, The, harmonize with scientific facts ... 199

Scotch, The, comparatively free from indigestion 138
" food of the poorer classes of ... 138
Schiaparelli on comets causing meteoric showers ... 181
Sea level, no such thing as a ... 165
Seals, transported from salt to fresh water ... 31
Sepoys, flesh eating ... 137
Sensational philosophers 78, 81
Shark, has little or no gills ... 116
Siam, ruined cities of Cambodia in ... 115
Simpson, Sir J. Y., on disease... 187
Sky, why is it blue? ... 82
" " " Prof. Tyndall's explanation ... 82
Snow flakes, cause of their beautiful forms ... 10
Solar heat, Prof. Tyndall says, is the true origin of glaciers... 109
Solent, the ship, a circular storm passed over ... 131
Sound generates heat, says Tyndall ... 102
" heard better on a winter's day than in summer ... 103
" a property of sympathy... 104
" Tyndall's new theory of... 105
" experiments at South Foreland, England, on... 105
Sorby, Dr., on water in granite 108
" on meteorites ... 180
Space, temperature of ... 71
Spectrum of colours, why seen 82
Spencer, Herbert, philosophy shaken ... 8
" based on a wrong foundation. ... 8
Spiders, mathematical knowledge of ... 34
Spontaneous generation, Blackwood's Magazine on ... 26
" the French Academy's prize for essay on... 24
" M. Pasteur on ... 24
" Dr. Child on ... 24
" Germ theory in opposition to ... 24
" fish produced by ... 27
" Mr. H. Charlton Bastian on ... 28
" insects produced by, through Mr. Crosse 20
" atomagnetism the cause of .. 29

St. Louis, United States, once a fever swamp ... 115
St. Paul's Cathedral, how long fifty organs would take to heat 103
Steam boiler explosions ... 47
" " " U. S. Commissioners on ... 48
" Com. Vanderbilt's Staten Island Steamer's 84
" caused by combustion between oxygen and hydrogen ... 51
" prevention of ... 51
Stomach, like a magnetic battery ... 189
Storms, how the inferior animal knows of the approach of ... 35
" cause of ... 127
" Sir John Herschell on 128
" Prof. Rogers on ... 128
" the magnetic curves of the earth, the cause of 128
" Maury on circular ... 130
" Description of a circular 131
" snow storm in Chicago 132
" wind, blow to a centre, not round one ... 131
" raise the tide ... 161
Sun, the, Professors Thomson and Tait on ... 70
" Sir John Herschell on the waste heat of ... 71
" new theory of ... 71
" inhabited ... 74
" not a furnace ... 76
" winds caused by ... 128
" forms glaciers ... 109
" repelling power of, on comets 174, 175
" influence of, on the tides ... 162
" Mairan on the extent of the atmosphere of ... 182
Sunlight, how caused ... 73
" is electricity ... 73
Swallows, instinct of ... 34

T.

Tables of food, Prof. Lyon Playfair's ... 134
Tails of comets, length of ... 176
" " cause of ... 176
Tait, Prof., on energy and the sun ... 69

Tantramer River, Bay of Fundy, tide rises two feet in three miles ... 165
Telegraph, arrangement of ... 94
" worked by grass ... 97
Tides, regularity of the ... 161
" in the Mediterranean ... 161
" in the Bay of Fundy, and cause of 161, 165
" high, at new and full moon ... 165
" storms raise the ... 166
" caused by pressure on the plane of the Ecliptic ... 164
" cause of variation of the 164
" said to be caused by attraction ... 164
" Lardner's theory of, untenable ... 164
" average rise of the ... 163
" follow the Meridian ... 163
" moon's influence on the, explained ... 164
Thomson, Sir William, on seed bearing meteors ... 12
" on origin of life ... 12
" on energy ... 69
" on how the sun's furnace is fed ... 70
" on electricity flowing 85
" on the internal fire of the earth ... 15
Thompson, Dr. Wyville, against Dr. Carpenter's theory of oceanic circulation ... 169
Thomson, Dr. Thos. on electricity ... 85
Trees, attract only material similar to themselves 7
" result of coating, with paint ... 16
" why they do not grow in winter ... 18
" composed of carbon ... 140
" formed of an aggregate of individuals ... 149
Tyndall, Prof. on Matter and Force at Dundee ... 2
" on dust and disease ... 25
" on heat as a mode of motion ... 48
" on Sir Humphrey Davy's ice experiment ... 54
" lectures on heat ... 54
" on the heat caused by the stoppage of the earth ... 61

INDEX.

Tyndall, on souls of fire ... 61
" on Dr. Thos. Young and the undulatory theory of light 66, 63
" on iron and lead bullets ... 60
" on the heat caused by the earth falling into the sun ... 61
" on waves of coloured light ... 79
" on electricity ... 86
" on magnetic curves 92, 93
" on sound ... 102
" his experiments at South Foreland ... 105
" new theory of sound ... 105
" on glaciers ... 109
" on the internal fire of earth ... 154

U.

Undulatory theory of light discovered by Dr. Young ... 66
" follows Emission into oblivion ... 68
" Tyndall on the stability of the ... 68
Uxmal, Central America ... 115

V.

Vacuum, how caused in the atmosphere ... 128
" causes wind ... 128
Vegetable life ... 11
Volcanoes, Dr. Mayer on the cause of ... 154
" Sir John Herschell on 155
" Prof. Mallet on ... 158
" atomagnetic theory of 157
" when apt to become dangerous ... 159

Volcanoes, Mount St. Helena, California, extinct 51, 158

W.

Water, more powerful than fire 107
" in granite ... 10
" of what composed ... 108
" manufactured in the atmosphere ... 109
" Tyndall's Forms of ... 109
" no air in ... 115
" Tyndall proves air to be in ... 115
" efficacy of hot, in disease ... 191
" danger of iced ... 192
Waves of coloured light, Prof. Tyndall on ... 79
Weather and weather prophets, Sir John Herschell on ... 128
Weather maps, American ... 132
Wells, Dr., theory of dew ... 121
" experiments with wool packs ... 121
" says radiation is the cause of dew ... 121
" on the cause of moonblindness ... 121
Wheat, becoming rye ... 23
Winds, Prof. Rogers on the cause of ... 128
" the sun the cause of ... 128
World, the end of the ... 46
Wrecker's light, faith in man like to the ... 130

Y.

Young, Dr. Thos., author of the Undulatory theory of light ... 66
" Tyndall's estimate of ... 66

www.ingramcontent.com/pod-product-compliance
Lightning Source LLC
Chambersburg PA
CBHW020807230426
43666CB00007B/898